CRACKLE AND FIZZ

Essential Communication and Pitching Skills for Scientists

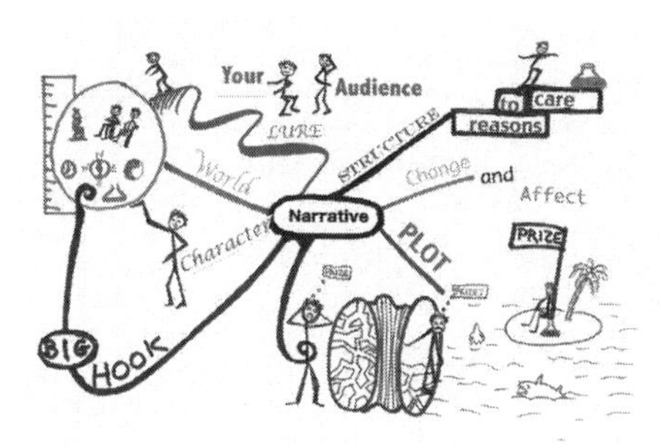

CRACKLE AND FIZZ

Essential Communication and Pitching Skills for Scientists

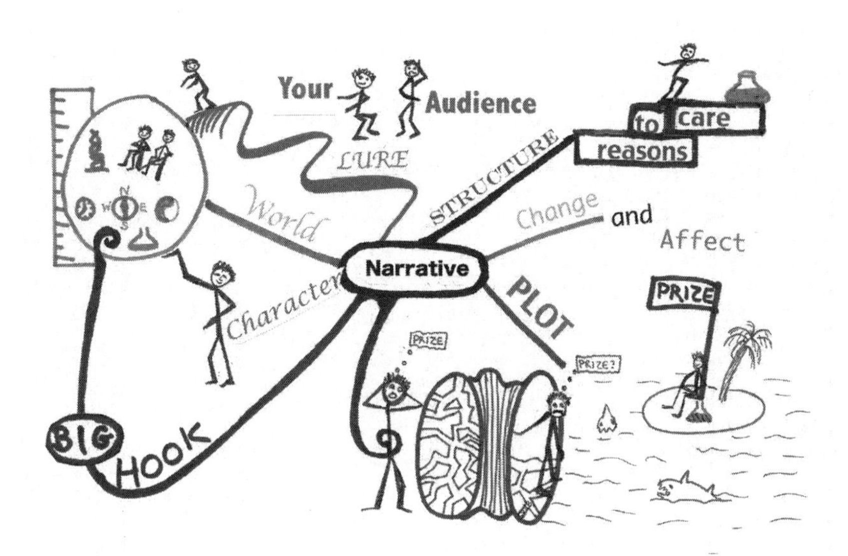

Caroline van den Brul

Creativity by Design, UK

Foreword by

Professor Nancy Rothwell FRS

ICP

Imperial College Press

Published by

Imperial College Press
57 Shelton Street
Covent Garden
London WC2H 9HE

Distributed by

World Scientific Publishing Co. Pte. Ltd.
5 Toh Tuck Link, Singapore 596224
USA office: 27 Warren Street, Suite 401-402, Hackensack, NJ 07601
UK office: 57 Shelton Street, Covent Garden, London WC2H 9HE

British Library Cataloguing-in-Publication Data
A catalogue record for this book is available from the British Library.

CRACKLE AND FIZZ
Essential Communication and Pitching Skills for Scientists

ISBN 978-1-78326-283-0
ISBN 978-1-78326-284-7 (pbk)

Typeset by Stallion Press
Email: enquiries@stallionpress.com

Printed in Singapore

Contents

Foreword

Escaping an angry iguana, being bitten by a lobster, being peed on by a monkey and avoiding frequent explosions were part of my best learning experiences in science communication. These encounters, and many more, were part of the *Royal Institution Christmas Lectures* that I gave in 1998. But such hazards were not why I learnt so much. That was due largely to the executive producer of the lectures, Caroline van den Brul.

I first met Caroline in 1979 when I was a young researcher and Caroline was making her first *Horizon* programme, called *The Fat in the Fire*, about my research with Mike Stock on brown fat. By 1998, Caroline was a senior executive producer for the BBC and a highly respected science communicator. During the making of the *Christmas Lectures*, Caroline advised and tutored me about understanding the audience, exciting and intriguing them, and, above all else, the importance of telling stories. Since then she has taught many junior and established, naïve and experienced scientists and non-scientists about communication.

Crackle and Fizz captures Caroline's vast expertise and experience in an engaging and informative format. It describes why you need to know your audience, how to captivate them and keep them interested, how to develop a plot and how tell a story in order to get your messages across. It is full of anecdotes and examples, tips and exercises and can be read from cover to cover or dipped in to regularly and kept as a reference. I shall certainly recommend it to students and colleagues.

Some might think that *Crackle and Fizz* is for those scientists who want to communicate with the public. Not so. Communication between scientists, even those within the same field, is critical to success. We have surely all read a beautifully written scientific article or grant proposal that is a pleasure to read and a delight to understand. I once read a grant application in physics which was so well written that, even as a biologist, I could follow it – or at least the author made me think I could. Not surprisingly it received full funding. *Crackle and Fizz* can help us all to communicate better.

I just wish I had been able to read it before the *Royal Institution Christmas Lectures*, which remain the most difficult and rewarding experience of my career.

Professor Dame Nancy Rothwell, FRS, FMedSci, FIBiol
President and Vice-Chancellor
University of Manchester

Introduction

In 2005, a charity which funds science research decided to run a brave but risky experiment. A number of their senior researchers were invited to an evening event; so too were a group of journalists. The idea was to bring the two groups together to begin to build relationships between those working in science and the media. The central activity of the evening was based on the principles of speed dating. The format was that each scientist would have a few minutes to explain their work to a journalist; a bell would signal time up, and the journalist would move on to hear the next researcher.

The evening was not a success. It was, by some accounts, little short of a disaster. Far from building relationships, the event created mild hostilities. The researchers complained that the journalists were arrogant, not interested in the science, only wanted soundbites. The journalists were frustrated that they had little idea what the researchers were talking about. The researchers made no concessions to them as a non-expert audience, and there was a full-scale communication breakdown. As a result, several people declared that they would not be coming to any more events like this one.

This minor crisis turned, as all good crises do, into an opportunity which the charity moved quickly to exploit. The science researchers' feedback was clear: they wanted to be able to communicate better with non-scientists and they asked for training in 'story' – a concept bandied about by the media professionals, but which seemed far removed from their research. These events turned into an opportunity for me, too.

I had recently retired from the BBC, and I was looking for a new challenge. After my chemistry degree I had made a career in the BBC. For nearly 30 years I made television programmes about science. From half-hour, degree-level biochemistry programmes at the Open University Productions, to 50-minute *Horizon* documentaries on subjects as various as obesity, trees and gender, as well as scores of shorter stories about the latest developments in research for popular science magazine programmes. I produced films and studio pieces on a huge variety of topics from monoclonal antibodies to fractal mathematics, from thermogenesis to a milking machine designed for the Lacaune ewes whose milk makes Roquefort cheese. I learned how to make science appealing and comprehensible to audiences not trained in science. I also learned, from the people I worked with, what it was like to be a practising scientist or engineer.

If there was one series which provided me with the best training for writing a book on science communication it was the five hour-long sets of annual *Royal Institution Christmas Lectures.* Their Audience is children.

I led the teams who helped preeminent scientists and mathematicians shape, prepare and present their work to young audiences. All of us on the production and technical teams, with our different backgrounds and knowledge, had to get to grips with subjects which are not easy to explain – quantum physics with Neil Johnson, mathematics with Ian Stewart FRS, the brain with Susan Greenfield, geology with James Jackson FRS, palaeontology with Simon Conway Morris FRS and homeostasis with Nancy Rothwell FRS. We not only had to work with the lecturers to find ways to explain their topics to an audience without any specialist knowledge, we also had to take into account that children would not sit quietly in their seats for an hour or more if they were not completely engaged in the action.

By extraordinary coincidence when I first went to work as a researcher at the BBC in the education department, my producer handed me a faded pink pamphlet. '*This has all you need to know,*' he said, '*about how to make television programmes.*' The pamphlet, *Advice to a Lecturer,*[1] was loaded with recommendations about how to lecture and how to present scientific subjects to an audience. There was no mention of television, for the simple reason that television had not been invented, but even so the general advice within this pamphlet was relevant. Its author was 21-year-old Michael Faraday – the brilliant chemist and experimentalist who in 1825, as Director of the Royal Institution, introduced what would become an annual tradition: the Christmas Lectures for children.

Many of the scientists I meet today remember being inspired by these televised lectures. For most, the inspiration never left them; they were motivated to seek a career in science. So how did we make difficult material understandable to children while keeping it truthful?

We followed in the tradition set out by Michael Faraday and did what he and others had worked out long ago. We used stories and narrative techniques to draw in our audience, keep them engaged and present them with ideas and information. To a small extent we educated them, we certainly entertained and inspired them, and above all we gave them memories.

It was an approach which became second nature to me. I had, without making too much of it, helped scientists, academics, researchers and television producers use narrative skills to communicate their ideas effectively for years. I had spent my entire professional life as a science communicator.

Yet, until shortly after the speed dating event, when the charity and I began to discuss the idea of a storytelling workshop to help their researchers, I had never sat down and thought about my craft with a scientific eye. It had not occurred to me to do so. Why would I? I knew how to do it, and mostly I had learned by doing, and by being in an environment where everyone else was doing it too.

With the charity's encouragement, I set out to construct a plan for a workshop to give researchers an insight into how they could engage and influence people with their work. I was clear that storytelling was critical. Communication is not only about exchanging information and making yourself understood, although it is both these things; it is primarily about creating an experience which people will remember; and remember in a way you would like them to. Stories can do this.

I set out to design an event which would allow scientists to rediscover what they already knew about stories and narrative. (I use both words loosely to include the content and how it is put together.) I would introduce them to eight critical ingredients: the Audience's needs, the Lure to grab the Audience's attention, the World where the story takes place, the Characters (often the storyteller in specialist communication) who inhabit the World, the Big Hook which raises a problem as a critical question that the Audience wants answered, the Plot which outlines the difficulties and choices in getting to the answers, the Structure giving the narrative a shape and the Audience reasons to care, and finally the Change and Affect: insights, surprises and new thoughts, which complete the story journey and give the audience a sense of intellectual and emotional fulfilment.

I drew mostly on my personal experience but I was also influenced by the ideas of others, giants in the field of story, who have written extensively about it: Robert McKee,[2] Christopher Booker,[3] Annette Simmons,[4] Stephen Denning.[5] As part of my training as a television editor I had been on one of McKee's famous screen writers' courses which have inspired not only those involved with Hollywood movies, but novelists as well. I went back to the source of his ideas: Aristotle,[6] and I read others too.

This book is a culmination of my personal experience of grappling with complex factual science and making it comprehensible and engaging for others. I have developed the content over several years working with scientists, engineers and academics in my narrative skills, communication skills and self-presentation workshops. Some of their stories are here too.

The book uses narrative, the currency of the creative industries, to show research scientists how to harness storytelling principles to make their complex or technical content both easier to communicate and more fulfilling for their audience.

The communication principles at the heart of this book, together with the exercises and examples will, I hope, inspire people to think about communicating their work, results and ideas in a different way, by using narrative. The book provides a framework to help scientists and other experts shape their content so that it *crackles and fizzes* with relevance for their intended audiences.

I would like to thank the hundreds of research fellows, laureates, scientists and engineers from industry and academia who have been on my workshops and motivated me to write this book. I hope that it triggers useful memories and helps others who read it to communicate widely their vitally important work.

Acknowledgements

I am indebted to my former colleague, one time editor of the BBC series *Tomorrow's World* and a dear friend, Dana Purvis, for her skilful and generous editorial support of this book. Her thoughtful suggestions and perceptive questions were invaluable in helping me rewrite the many drafts, and her creativity was instrumental in sparking the idea for the title.

The BBC science department was where I learned and practised my craft of storytelling. I am grateful to my many colleagues for their ideas, insights and ambition over the years. They provided a vibrant and stimulating workplace in which to grow and learn. Particular thanks are due to all the scientists and engineers who allowed us into their labs and offices. We were privileged to be able to help them share a joint passion and commitment for science with a wider public.

Thanks are also due to the public engagement team at the Wellcome Trust under Dr Dan Glaser, and to the Wellcome Trust research fellows who were the catalyst for this book. I would like to thank Alom Shaha – another former BBC colleague – for recommending me to Rachel Hillman at the Trust and to Chris Stock, with whom I developed the first pilot narrative skills workshops.

Thanks too to Kirsty Jones, Chloe Sheppard and Verena Collins-Magee for their solid support over the years.

I further developed the workshops for scientists – to include presentation skills – with the support of the Foundation for Polish Science SKILLS project; warm thanks are due to Katarzyna Pronobis and Alexandra Krypa as well as their teams of laureates and scholars.

I had been thinking about writing a book to help scientists with their communication skills for about five years, but, like many people with such thoughts, I had done little about it. I would like to thank Dr KK Phua, chairman and editor-in-chief of World Scientific Press, for finally persuading me to write it, and for introducing me to his colleagues from Imperial College Press – all in the space of a lunchtime at the London Book Fair in 2012.

The manuscript and any errors (which I hope are few) are mine. I would, however, like to thank Isobel Lawrence, Serge Mostowy, Simon Rayner, Gerald van den Brul and Charlotte Houghton for their helpful comments on individual chapters. Thanks too to Geoff Moggridge, Simon Rayner and Joanne Heng for their pitch contributions. Thanks are also due to the editorial staff at Imperial College Press: Kellye Curtis, commissioning editor, Tom Stottor, the copy-editor and particularly my editor Tasha D'Cruz who supported me through the whole process with patience and professionalism. I would also like to thank Janice Rayment for her index.

Special thanks to Nancy Rothwell for her foreword, for reading the manuscript and for her friendship over many years. Finally, I owe more than I can express in words to my husband, David Hague, who always believed, especially when I doubted, that I would finish the book, and who kept me going with unstinting support, endless cups of tea, and love. Thank you.

1

Your Audience: the person or people whose full engagement you need

> 'A lecturer may consider his audience as being polite or vulgar (terms I wish you understand according to Shuffleton's new Dictionary) learned or unlearned (with respect to the subject), listeners or gazers. Polite company expect to be entertained not only by the subject of the Lecture but by the manner of the Lecturer; they look for respect, for language consonant to their dignity and ideas on a level with their own. The vulgar (that is to say, in general, those who will take the trouble of thinking) and the bees of business wish for something that they can comprehend. This may be deep and elaborate for the learned, but, for those who are as yet tyros and unacquainted with the subject, must be simple and plain. Lastly, listeners expect reason and sense, whilst gazers only require a succession of words.'
>
> Michael Faraday[1]

Walter Marshall, a former chairman of the UK Atomic Energy Authority, used to tell a story about the occasion Margaret Thatcher, as leader of the UK Conservative party, visited Harwell. She was given a tour of the DIDO nuclear reactor and taken to see an experiment involving the neutron scattering of proteins. The researcher told her what he was doing but, in the process of doing so, simplified the story to such an extent that what he said was not actually true.

Margaret Thatcher interrupted him and snapped: 'I did research on proteins and what you have just said is rubbish. I think I can better use my time talking to someone else!' And she walked out. At the end of the day, Airey Neave MP, who was accompanying her, said: 'You were very cruel to that man, you know, because he was only trying to make it simple. You ought to go back and make friends with him again.'

So Mrs Thatcher went back, the researcher explained his experiment in its full complexity and, of course, she got lost. Walter Marshall used the story to illustrate how Mrs Thatcher listened and could tell when she was told a false story; but I retell it because it shows how difficult it is for scientists to pitch explanations of their complex science at a level which is both understood by their Audience and also true. For scientists or experts who aspire to communicate with people outside their own narrow specialist area, it is a skill which must be learned.

> The Audience is the compass which should guide the shape and choice of content in an effective piece of communication.

This book takes a particular approach to communication. It focuses on using stories and a narrative structure to convey ideas, information and meaning to an Audience. The process of constructing a narrative is dissected into eight separate chapters, each covering a different narrative ingredient. In each chapter we look at the part the ingredient plays in narrative and how it can be made relevant to the Audience.

In this chapter, my intention is to provide an overview of the scope of the book and look at some of the underlying theory about the Audience and 'Audience needs'. This theory explains what motivates people and why they behave in particular ways. While it is not necessary to understand the details of the theory, it helps to know the general principles, because they account for Audience receptiveness for, and reactions to, new ideas and new information.

There is also an example showing how professional communicators, who are paid to get a message across to a particular Audience, go about doing so.

Finally there are some questions to ask yourself each time you need to plan a presentation or poster, to help you focus on your specific Audience.

There are three basic rules for successful communication:

Connect with your Audience

How? Know their needs. Use your understanding of who they are; what they know and what they do not know; and what they expect and want. This is discussed further in the *Lure, World* and *Make your Pitch CRACKLE* chapters.

Stimulate your Audience

How? Give them a memorable experience – one that takes place in their minds; create stories. This is explored in detail in every chapter.

Be Understood

How? Know *clearly* what you want to convey; know your Audience's limitations. The chapters on *Change and Affect*, *World*, *Plot*, and *Structure* discuss this in more detail.

Connect with your Audience

Understand their needs

Consider yourself as the Audience in these seven different situations. What are your needs and expectations from the engagement? What might be going through your mind as you get ready to listen or read? Use the full range of desires which you hope and expect the communication to satisfy.

1. Going to hear a lecture from a renowned expert about an interest of yours.
2. Receiving a pitch asking for money, resources or time, which you control.
3. Reading a draft of a press release about your work.
4. Attending a first departmental meeting with a new boss.
5. Attending a project meeting to be told about significant budget cuts.
6. Reading a paper that is relevant to your research.
7. Listening to an after dinner speaker at a conference.

If your list is like mine, your expectations will be split into two distinct groups. One group addresses my intellectual needs around the facts, the ideas and whether the evidence stacks up. The other group is linked to my feelings and emotions. How the facts, ideas and evidence might affect me personally: make me curious, concerned, proud, or perhaps give me the spark of an idea.

In formal science communications it is usually easier to make assumptions about the intellectual needs of the Audience than about their emotional needs. Acknowledging both, however, increases the likelihood that you will connect more fully with the Audience and be able to find effective ways to stimulate their interest. There will be more on this in the next chapter, *Change and Affect.*

Here are my assumptions about my hopes and expectations, my emotional and intellectual needs in each of the seven situations above:

1. Going to hear a lecture.

Intellectual needs: new information, deepening of knowledge, gleaning new insights, building a mastery of the subject. Observing and analysing the speaker's methods, style.

Emotional needs: prestige (status) associated with attending an expert's lecture, a glow which comes from belonging to a community of interest, anticipation of meeting colleagues, enjoyment, pleasure.

2. Receiving a pitch.

Intellectual needs: how is the content meeting the brief for the project? Where is it not? Is this idea novel, interesting, feasible? When will it deliver? How much will it cost? Is it good value for money?

Emotional needs: is this someone I can trust? Do I like this person and the way they go about their work? Do I admire or share their values? Can I see myself, my team, working with them?
(Constructing a pitch requires a particular kind of narrative which is discussed in detail in the final chapter, *Make your Pitch CRACKLE.*)

3. Reading a draft of a press release.

Intellectual needs: is the press release accurate in every detail? Is it likely to attract media attention?

Emotional needs: recognition for what I have achieved, nervousness with how my community of peers will view the release. Curiosity – who might respond to the release?

4. Attending a first departmental meeting with a new boss.

Intellectual needs: what has this person achieved? What is his background and interests? What will change with the arrival of the new boss?

Emotional needs: sadness – departure of previous boss. Reassurance – will we get on? Concern, nervousness, curiosity – who is this person, what are their likes, dislikes? What will change – will I lose anything I value?

5. Attending a project meeting about budget cuts.

Intellectual needs: what is happening? Why? When? How? Who will be involved?

Emotional needs: reassurance – are my job and income safe? Concerns or fear for self-esteem and status: what will I lose? If there is a break up of a work group how will the community and my relationships be affected?
(Giving information to colleagues about matters which will change their working conditions is likely to be a narrative of many

instalments telling an unfolding story. In each episode you should find ways to acknowledge the needs of the Audience, even though you may not be in a position to satisfy those needs.)

6. **Reading a paper that is relevant to one's research.**

Intellectual needs: is the methodology sound? Is the proposed theory feasible? Do I agree? Is the case for this proposition based on good evidence? What other explanations might there be?

Emotional needs: excitement, curiosity – how does this relate to my work? Where does it fit or compete?

7. **Listening to an after dinner speaker.**

Intellectual needs: possibly none!

Emotional needs: to be entertained, to laugh, to be amongst fellow human beings having a good time.

These are my thoughts; yours might be different. The emotional needs of the Audience are not trivial. They are the primary means for connecting with an Audience; and they affect how an Audience will react to your communication or pitch. There is more about this in the *Change and Affect* and *Make your Pitch CRACKLE* chapters.

> Audiences will connect with a communication when it speaks to both their intellectual and emotional needs.

Audience needs are 100% self-centred

Your partner, parents and supervisor apart, Audiences – whether a public Audience, a funding committee or your CEO – are not primarily concerned with your emotions: how you are feeling, whether you are nervous or not. They are certainly not interested in what you think they *should* know – at least not until you have shown them why it is worth knowing. Audiences are focused on their own interests and agendas. They mostly want to know: 'How is this relevant to me?' or more directly: 'What's in it for me?'

Some theory: Audiences have basic human needs which are universal

Abraham Maslow[2] was a psychologist interested in what motivates people. What drives us to do certain things, to avoid others, to behave in a particular way? What makes us value life? What influences our desire to learn, to bungee jump, to share our last piece of chocolate? Why do we join clubs and get married? What drives us to want to want to do science, and to tell others about it? In communication it is helpful to look at the questions from the Audience's point of view: what motivates us to want to read an article, look closely at a poster or listen to someone telling us about science?

Maslow developed a model of human needs: those internal forces which drive our thoughts, actions and behaviours, and those of every other person on the planet. His model is known as Maslow's hierarchy of needs. He identified the most basic human needs as the physiological and biological needs for survival: air, water, food and sleep. If we are trapped under water, even for a short time, our sole desire is to find air; and quickly.

Air Water Food
Sleep Sex

Biological needs

We may have deep concerns about our work, relationships or finances, but when we lack air, water, food or sleep, these concerns are obliterated; our motivation is utterly focused on getting what we need to survive.

When the basic physiological needs are satisfied, the overriding concern for humans turns to shelter; people take actions to feel safe and secure.

Security **Stability** Shelter

Safety needs

According to Maslow, once people feel secure they look to satisfying their social needs. Having a family, relationships, friendship and love are important human 'belonging' needs which people seek to fulfill the world over.

Community Belonging Involvement
Friendship **Love**

Belonging needs

In Maslow's next 'level', people are driven by their need for esteem. Once people feel secure and are part of a community, they develop an overwhelming desire to achieve, to be respected, and to have status and prestige within the group.

People take positive action and behave in ways which will help them to achieve these esteem needs as they do for their other human needs. Although Maslow defined human needs as a hierarchy, many other psychologists have argued that the hierarchy is not a rigid form. People's needs change with circumstances and situation.

Mastery Achievement Self-Esteem
Status **Prestige** *Respect* Self Respect

Esteem needs

The biological, safety and esteem needs were defined by Maslow as deficiency needs because they arise from something which is missing from an individual's experience. If the needs are not met and their lack persists, the individual may become depressed or anxious and unhappy.

There is another set of human needs which Maslow recognised as separate from deficiency needs. These are growth needs or 'being' needs, often referred to as the need for self-actualisation. That is the drive to develop to one's full potential, to satisfy a higher purpose in life, to find a reason for one's own existence.

Self-Actualisation
Intellectual Fulfillment
Life's Purpose
Growth needs

Specialists and experts, particularly in Western societies, are often fortunate to be able to pursue self-actualisation as a full-time occupation. Work is, for the most part, pleasurable and fulfilling. Abstruse knowledge can be legitimately pursued for its own sake, but its communication presents particular challenges.

If you consider Maslow's hierarchy of needs as a basis for understanding human motivation, it follows that the Audience's needs may be very different from the speaker's or writer's needs. Audience needs may not include growth needs at all. Needs for esteem, respect, to be part of a community, for example, may be as pressing, in some situations, as their need for intellectual fulfilment and self-actualisation.

> 'The most mature person in any social setting is the one who is most adaptable to other people's needs.'
>
> John Dewey

Conversely, speakers also have needs. This is how I recall being introduced to five women in Mississippi when I was making a BBC documentary, *Eating Earth*[3] – or as the women called it, 'eating dirt'.

> 'Good morning, my name is Professor Archibald Smith Jr. I am the Head of the Department of Anthropology at 'a University'. I have brought this lady here to your trailer today, not because you are black and of a low socioeconomic group, but because she is researching the female phenomenon of geophagy.'

I was surprised by the professor's clear need for esteem. It appeared more important to him than his need to connect with his Audience.

Stimulate your Audience

Science researchers' communication is rarely without audio-visual aids. Those giving presentations frequently show PowerPoint slides and pictures; some draw on whiteboards or flip charts; some create demonstrations or present models for people to handle. Some speakers are able to modulate their voices, vary their volume and tone, even use recorded sounds to play in their presentations. They are able to show their passion for their subject in their body language, their gestures and manner of delivery. Poster designers use stimulating visual images and clear type. Using a range of material to engage the visual and auditory senses of an Audience is normal practice and can breathe life into any form of communication.

There is, however, a particularly powerful presentation technique which can apply to all communications; posters, presentations and written pieces alike. It does not depend on physical or audio-visual material, although these can enhance the experience. The technique is, as you have probably guessed, presenting the content as a story.

Well-told stories have the capacity to stimulate all the senses because they create mental experiences for the Audience. If you are able to consider your science or research project as a story and use narrative techniques to let it unfold in the Audience's head, you have the best opportunity to communicate your ideas.

Stories are experiences which take place in the mind. To tell stories well requires an ability to transport the Audience into the World where the story takes place. (Of which, more in the *World* chapter.)

My former colleagues in radio always like to remind their television cousins that the pictures are much better on radio. A researcher who has a story to tell needs first to see it, in all of its detail, unfolding in his own mind.

> 'To the thinking soul, images serve as if they were the contents of perception. That is why the soul never thinks without a picture.'
>
> Aristotle[4]

The skill is in making choices of what to present to the Audience, of how to present it and in what order. Another skill is to be able to describe vividly the details that will bring the science to life; show what it is like to be there, within the cell, on the surface of the smart material, in the heart of the lab or out in the field. The narrator, the author or poster designer needs to provide the ingredients and put them together vividly for the Audience. These are the subjects of the following chapters.

Be Understood

Below are some words in random order. The task, which has a purpose, is to look at the words and memorise them.

> are are because because dare dare, Seneca is is not not not difficult difficult. do we we that that It things it things do

What is going through your mind as you look at the words?
What are you feeling as you look at them?
What tactics would you adopt to be able to reproduce these words in five minutes time?

The words, on their own, are simple, comprehensible and unambiguous. They are not, however, placed in an order which makes any sense. It is hard not to find the task irritating and a complete waste of time and energy. When I ask people to do a similar exercise in my workshops they wonder whether, with some reordering, the words could be put together to make sense. As for memorising it, it takes effort, but most people find strategies to complete the task. They use tried and tested methods of remembering random facts: exploiting perceived patterns, creating a mnemonic and the fallback – learning by rote.

When I have done this exercise in China, the participants, without exception, reorder the sentence before they attempt to memorise it. They invariably write up the quotation from Seneca as it is normally presented:

> 'It is not because things are difficult that we do not dare, it is because we do not dare that things are difficult.'

The point of the exercise, of course, is to demonstrate that if a sentence, an idea, a diagram, a formula, a research paper, an account or a story is not understood, it takes an enormous amount of effort to memorise it. It becomes almost impossible to recall later. For people to remember, they need to understand. The task for experts who deal with complex factual material is a difficult but essential one: to be comprehensible to Audiences.

First – Understand for yourself

It may seem self-evident, but the first step in planning to convey an idea or complex information to an Audience is to make sure that you fully understand it yourself. What is it that you want to convey to this Audience? What is your intention? What are your ideas? Are they clear to you? Can you express them in a way that makes sense to you?

Stephen King, the thriller writer, in his book *On Writing*[5] told of some advice he had been given very early in his career when he was working as a journalist. It was to write the first draft of a story with the door closed. The first draft is when you – as a writer or designer – are working out the story in your own head. The advice for writing the second draft? To have the door open. Once you understand yourself what you want to say, you open the door and rewrite it with the Audience in mind.

For this book, my own approach to having the door open was to shape the content for two individuals, whose photos I have pinned above my computer. Both were, at first, healthily sceptical about using narrative to communicate science. As I go through the rewriting phases, I ask myself: 'Have I made this relevant and stimulating for these two researchers? If not, how can I alter it to make it *crackle and fizz* for them?'

Second – Be understood by your Audience

Once your ideas are clear in your own head, and only then, think about the Audience. How might you use their knowledge and experience to convey your ideas? This becomes easier if you are aware of what the Audience already knows and if you are able to create a story from your material.

Some more theory: Audiences understand through association

If a page of words representing different objects is presented to someone who cannot read, the content, like the Seneca quotation, would mean nothing. Add a recognisable picture of the object and the meaning will be understood. It is a simple process of association which forms the basis of how we learn. Roger G. Schank is an Artificial Intelligence expert. He believes that stories are the *only way we learn and remember.* Stories provide our brains with frameworks for understanding the world and how it works. Unless we create a story for ourselves when we learn something new, we will be unable to recall it later. His thesis is presented in his book *Tell me a Story.*[6]

Schank argues that the most intelligent people are those who are able to remember the most stories. The way that people remember stories is how they link them with their experience, what they already know. Schank provides examples of how individuals 'index' stories in their brain. When we come across any new information or experience – from a train timetable or a diagram of the cell, to a book, film or a first day in a lab – our senses engage and so does our brain. The brain actively dissects, sorts and stores every nugget of information, every sensual detail we see, hear, smell, touch or feel. Crucially, it indexes and cross references all the new details with information we already have stored in the brain from previous experiences. This not only applies to the informational content, or what we heard of it, but also to the context in which we experienced the story, the emotions we felt and any consequences or puzzles which we linked to the story. The more connections you make, Schank argues, the more intelligent you are. This thorough sorting of all the different kinds of stimulus which are associated with each experience is really important in how we learn, and how we remember. Later, if we come across or conjure up any one of those stimuli, it can act as a trigger to recall the original story.

The more ways that our experiences can be indexed, the more triggers they will have for memory and recall later. Schank argues that the stories you tell yourself (or others) about any experience – however fleeting – and how you index them relate to your intelligence and ability to make connections between ideas and experiences. New information will be remembered by someone only if it links to one or more pre-existing mental frameworks in their mind. The more frameworks you can trigger with your communication, and the more diverse they are, the more memorable they will be.

A consequence of this is that if you want people to remember a story, you need to communicate it so that it connects to something that people already know. In other words, draw on metaphors or analogies which are relevant to your Audience. They can help you connect, stimulate and make you comprehensible to the Audience. (Of which, more in following chapters and in detail in the *World* chapter.)

Experts and Audiences

A corollary of having scientific expertise, like any other expertise is that experts have many more frameworks for understanding new ideas in their area of expertise than other people.

An oncologist, for instance, will have seen hundreds of cases of people with cancer. Each person's history and disease will be slightly different. These differences will be stored within her memory in a number of different frameworks. When a new patient is seen she will be able to integrate the complex information about their disease from their history, symptoms and test results. Some of the evidence may be uncertain, contradictory or ambiguous. Yet experts are better than the rest of us at dealing with complexity, uncertainty and ambiguity. They have the capacity to assimilate the information, integrate it with all of their previous experience and knowledge and to make decisions; or, in the case of the oncologist, recommendations on one or more courses of action. Non-experts are simply not able to do this with any confidence; and experts are only able to deal with complexity and uncertainty in their own area of expertise.[7]

Experts can create understanding through stories

Doctors create narratives to help patients understand complex material which may be uncertain and ambiguous. Other experts do the same. Just as doctors are able to create a story or a number of alternative stories to allow patients to understand their situation, they will similarly require the help of a lawyer to understand equally complex arguments in a legal arena.

Bankers from the world's richest countries illustrate the experts' domain of knowledge perfectly, but also provide a cautionary tale. Bankers' expertise gave them insights into the uncertainties and complexity of the global financial system. They were able to handle the complexity of different markets and to make their decisions accordingly. They explained their subsequent actions through stories: the ones they told each other, and the ones they told the public and politicians. Their decisions made them hugely wealthy at the expense of those who were not experts, who did not understand what was going on and who did not question the stories.

Stories create meaning for an Audience and it behoves the expert to be truthful in how they use stories – or risk losing the public's trust.

In any communication, the Audience will interpret its meaning through their own prism of understanding, which might be very different from yours. If experts are presenting to fellow experts they can safely assume that their Audience will be able to assimilate uncertainty, complexity and some ambiguity as part of the natural order of things. This is less likely with a more general Audience. The story needs to be carefully constructed, so any ambiguities or uncertainties are highlighted for what they are.

Risk

Risk is a particular form of complexity and uncertainty. Those scholars with training in mathematics and statistics are able to interpret and understand the significance of experimental results. They are able to compare and contrast situations through a prism of probabilities. When results appear which seem, to the untrained eye, to be counter intuitive, statisticians quite rightly trust the numbers. Innumerate Audiences do not. They trust their own personal experience and need clear guidance to help them understand the meaning of reported observations which could be accounted for by dozens of different possible explanations.

In many areas of life where complex decisions need to be made, such as governmental policies on climate change, human embryo research or space programmes, the knowledge provided by science is never the only consideration. Some well-publicised disasters in communication in the 1990s – the most notorious being the BSE outbreak in the UK, where the story from the Government was that there was no problem eating beef, when in reality, the science was showing there might be – created a crisis for British science. Since then, the scientific community in the

UK has become savvier about the need to tell a consistent story that public Audiences can understand. The example[8] which is often used to illustrate the improvement is the management of the communication around the science story which helped the Human Fertilisation Bill come into law. The communication of the big science stories in today's world requires a co-ordinated, managed communications offensive which is outside the scope of this book.

How the Professionals Do It – An Example

Advertisers are paid to communicate with Audiences. They convey ideas which are not the kind of complex ones which scientists work with. However, their approach is worth looking at for the ideas it might provoke. Advertisers pay a lot of attention to understanding their Audience needs. These needs provide them with a story, an insight, a way to connect with the target Audience, to stimulate their interest and make their ideas understood.

Let us look at an example: a challenge which was posed some years ago by the Central Office of Information (COI) in the UK. Anecdote backed up by research showed that fathers were not reading with their children. The COI wanted to produce an advertisement to persuade them to do so. Trying to persuade anyone to do something they do not want to do is tricky. It is the job of advertising planners to find a way to make the target group, in this case fathers, actually want to pay attention. Their job is to gather all the research on the target group: their likes, dislikes, behaviours and habits and any observations they and others have made. They try to deduce what all the different bits of information reveal about the group, beyond what has been stated before. That is, they are required to come up with some fresh insight which could give the advertisers ideas to help them fulfil their brief: in this case, to get fathers to read to their children.

They needed to find a way to connect, stimulate and get a message across to fathers so that they changed their behaviour. No mean feat.

The planners looked at this challenge and did some research with the fathers of young children. The reasons fathers gave for not reading with their children were, at face value, completely valid. They are ones which you could probably list just by thinking about it: fathers were at work all day and when they returned home, the children were usually in bed. So what about the weekends? The weekends, they argued, were filled with games, kids' parties, the weekly shop and other family outings. Not the time for sitting down to read a book.

The planners' observation matched the fathers' reports: they got home late and at weekends they spent time with the children but mostly out of the house. The planners understood that the fathers wanted to read to their children, but simply could not see how it could be done. The planners' insight was to understand and then *challenge the fathers' assumption* that reading with a child had to be done at home from a book. The insight was that 'reading is everywhere, not just in books'. This seems obvious once it is pointed out. Insights usually do. Yet this revelation guided their decisions about what to include in the advertisement. The result was a warm, inspiring film which gave the fathers ideas about different ways they could read with their children. The film featured stories of intimate moments with fathers and their small sons or daughters on trains, at football games, at the supermarket checkout, intently reading aloud from rail tickets, match programmes and the back of cereal packets.

An Insight

The capacity to gain an accurate and deep intuitive understanding of a person or a thing.

An insight is 'a not yet obvious discovery'.

An insight is an eye-opener, it is realised as an 'Ah ha' moment.

The skill of advertisers is finding ways to sell ideas. They often do this through presenting Audiences with a different way to look at something familiar. They challenge assumptions about the Audience and ask what later appear to be obvious questions. This is a skill which scientists have too. They may not ask the: 'Why?' and 'What if?' questions of Audiences, but they ask them in their research.

My final example brings an inquiry into Audience behaviour to the sphere of mainstream medical research. Professor Peter Rothwell works as a cardiologist at the John Radcliffe Hospital in Oxford. In the late 1990s he noticed that many of his patients, who had been referred by their GP, missed their appointments. These patients had been referred for further investigation following a transient ischaemic attack or suffering stroke-like symptoms. Peter Rothwell did not assume they had forgotten their appointment or had better things to do. He investigated. He asked the 'Why?' question and discovered that almost all the non-attenders at the clinic had been admitted to hospital. They had suffered a major stroke following their 'warning attack'.

This knowledge led to further research which showed that the risk of those with minor stroke events having a further event in the first week was about one in ten – much higher than was previously thought. As a result of this simple question about the behaviour of a group of patients, the clinical management of patients with minor strokes and transient ischaemic attacks began to change.[9]

If you were to go about researching your Audience in the way advertising planners do it, it would be time consuming and costly. However, there is much you can do simply by asking the right questions. When you come to look at the questions below and begin to develop ideas for your own communications, think about your own previous experience as a member of an Audience. For every good experience as a member of the Audience, consider: what made it so good? For every bad experience, go further. Ask what would have made it better?

Similarly, ask the same questions if you have been a speaker, an author or poster designer.

Questions and Ideas to Help Stimulate Ideas for Thinking About your Audience

Here are some questions which are designed to stimulate you to think about your Audience, their needs and the ways you might present your content. There are many more questions to help you focus your communication on your specific Audience in following chapters.

1. Whose attention do you want for your piece of communication?
   - Is it an individual, a small group or a large congregation?
   - Are they your peers, a senior group of researchers, academics from within the same discipline, the same sub-speciality?
   - Are they gatekeepers, those who make decisions about the allocation of resources? Do they decide who is given time to speak at conferences, who is allowed to display posters, or whose research is funded?
   - Are they journalists, public engagement specialists or the public?

Of course, the larger the congregation, the more important the communication is to you, the more prepared your content needs to be. Preparation may involve practice and rehearsals to make sure that your story connects, stimulates and will be understood by your Audience.

2. For your relevant Audience, list your assumptions of their expectations.
   - What are their preconceptions and assumptions *about you and your content*?
   - Can you design something to address, or perhaps challenge, those assumptions?
   - Can you find out more about the nature of the Audience by asking people who represent them, or who are familiar with them?

3. What needs do you think your Audience will have?
   List their motivations for looking at your poster, coming to your presentation, reading your article.
   — What are their potential intellectual and emotional expectations which you plan to keep in mind when you are constructing your content?

4. Consider your previous experiences as the Audience for a similar communication challenge.
   — What was your best experience and your worst? Think in detail why this was so in each case.
   — What did the author or presenter do to make the experience worthwhile, or not, for you?
   — How did they do it? Think about content and delivery.
   — Can you learn from their example?

Finally: try keeping a notebook or file of communication examples which strike you as effective ways of connecting with, and maintaining the interest of, an Audience.

Change and Affect: gives communication its meaning and impact

2

> 'I sit through many conference presentations and I understand every word that the speakers say, but I don't know what they are saying or why they are telling me'.
>
> PhD student in a narrative skills workshop

At school for our religious education homework we were expected to read sections of the Bible and write a précis of them in our exercise books. The sections were usually a single story which needed to be read, digested and summarised – an early lesson in storytelling. One of my fellow pupils, Maggie, who has since gone onto great things, used to try to use fewer words in her précis than anyone else in the class. I used to look forward to the homework simply to hear what Maggie would come up with. Her summary which I remember the most clearly – even after nearly 50 years – is that of the story from the Book of Samuel (1:17). It is the narrative of the epic battle between the Philistines and the Israelites which, in the end, is fought between two people: a young, slight shepherd boy, David, who has a staff, a sling and five stones; and Goliath, a Philistine giant, protected with heavy armour and carrying a shield. Maggie summarised the chapter in six words: 'David killed Goliath with a stone.' Our long suffering teacher sat with a pained expression after hearing this and, as every week, told Maggie that her homework was inadequate. As always, a discussion ensued as to why. It was, after all, a brief statement of what had happened; of what had Changed. The most important details were included: David victorious; Goliath slain; the weapon a stone. We debated how much detail should be put into a summary for it to be considered 'adequate'. We were only 11 and we missed the point, or at least I did. If you do not know who the two characters are, and if you have no idea why you should care about them, then the sentence means nothing to you. It will soon be forgotten.

If you already know about David and Goliath and understand who and what they represent, then the sentence is shorthand for the whole story and its underlying meaning.

A summary conveys meaning only when it includes both a Change and the Affects of the Change.

In the *Audience* chapter, I argued that communication needs to engage both the intellect and emotions of an Audience. In this chapter I shall build on that idea and show examples illustrating how the meaning of a communication is defined by the Change and Affect it creates for the Audience.

Stories are concerned with transformation. In stories something Changes to create an emotion or series of emotions in the Audience. The Change has to resonate with the Audience to generate an Affect; a feeling, a reaction or an insight. A game of football, a seminar, a photograph, an experiment, a scientific paper, a set of petri dishes recently covered in mould. All of these have the potential to become stories.

We shall consider these two defining characteristics of narrative to clarify the purpose of any communication. What is the Change I want to reveal to my Audience? What Affect do I want that Change to have upon them?

We shall also look briefly at how to harness the Change and Affect of familiar stories such as fables and classic tales by using them to enrich science communications. We shall demonstrate how a technical story can be told so that some of its meaning resonates with the Change and Affect of a different story, one within the Audience's experience.

Change and Affect – Entwined

When you make a contribution to a discussion, give a paper, present a poster or a talk at a meeting, you have the intention of giving your Audience information which will be useful to them. The best communicators create an emotional experience where the Audience is guided to think about what they are hearing or reading in ways they have not done before. They are doing more than simply providing a neutral account of the facts. They are providing an interesting frame through which to view or think about a subject. They do so through a narrative which effectively takes the Audience on an emotional journey. We are not talking roller-coaster emotions of extreme joy or fear, anger or surprise, but a wider variety, including anxiety, amusement, optimism, relief, frustration, pity, apprehension; indeed any visceral feeling or emotion, however mildly or strongly felt. Revealed over the course of a presentation or paper, Change and Affect make for a fulfilling experience for listeners and readers; they create meaning, they *crackle* the intellect, they make emotions *fizz*.

Figure 2.1 is a photograph[1] of two men with greyish hair in a largish room – one is sitting, the other is standing. Even if you do not know who the two characters are, it would be clear from the image that the individual sitting down has the higher status. This single image *crackles* and *fizzes* with meaning. It symbolises a monumental Change which has deep meaning for millions of people. The election of Barack Obama, with his part-African ancestry, as President of the USA was a story which resonated around the world. To each one of us who look at the picture, there will be Affects: a stirring of thoughts, memories and feelings. Not all will be profound or intense and they will not be the same for all of us. For every Change some things are gained and some are lost. It is the interpretation of the overall losses and gains and the emotions that these create, which give stories their power. They are how stories work for

Figure 2.1.

each individual. When my overall interpretation is shared by you, we feel some connection and we might say we are 'on the same wavelength'. When dozens, hundreds, thousands or millions of people interpret the Changes represented by a story in the same way they are bound together in a community of thought: a culture.

The election of Barack Obama as the 44th President of the USA was interpreted as a monumental Change by people beyond America. The overt Change was a new President at the White House, a Democrat, a black man. The meaning of that Change was made significant by its Affect in a broader context: an individual from a minority group, once segregated, oppressed and powerless, elected by his fellow citizens to the most powerful position in the land. It is a narrative which gave hope, optimism and a sense of triumph to millions.

I was visiting Grenada, part of the British Commonwealth, two years after President Obama was elected, and photographs of him and his family were everywhere: on the back of cars and vans, in cafés, in people's houses, in shops, in the street. One day I asked if

President Obama had visited Grenada (as I thought that might explain the vast number of photographs). I was told, in a voice brimming with emotion: 'No Mam! But he is a black man, and he is President of the United States of America!' The Affect of the Change on black people, and minority groups around the world was profound. Its intensity was at once cultural and personal. The significance of the Change is interpreted through the prism of each individual's prior knowledge and experience.

This applies to science communication too

Specialists with expert knowledge will be part of a subculture, with a shared set of stories, each offering some insights, some Change in their specialist World view which is relevant to the community. The Change in these stories may not mean anything to outsiders, but to the group, they represent something of importance which binds them in a common understanding.

If you are unfamiliar with the subculture of science, Figure 2.2[2] may not mean anything to you. If you have studied science to any level you might recognise it as something to do with DNA. You may hazard a guess, correctly, that it represents the first sketch of the structure of DNA which was brilliantly established by James Watson and Frances Crick. (The original published drawing was made by Odile Crick as neither Watson nor Frances Crick admitted to being able to draw.) This picture can generate powerful emotions amongst scientists who recognise the drawing and its significance.

For some it sends a tingle down the spine. To them it represents a transformation in scientific understanding of genetic inheritance. This deceptively simple sketch explains one of the most important structures in biology, an insight into how life is reproduced. For those who see how the double helix holds the ingredients of life in two parallel strings of precious molecules, it inspires ideas for future experimentation.

When I have shown this image to some female scientists however, their emotions have been quite different. For some women, this Watson and Crick sketch generates strong feelings of injustice. The World in which they view the drawing is not from the point of view of the

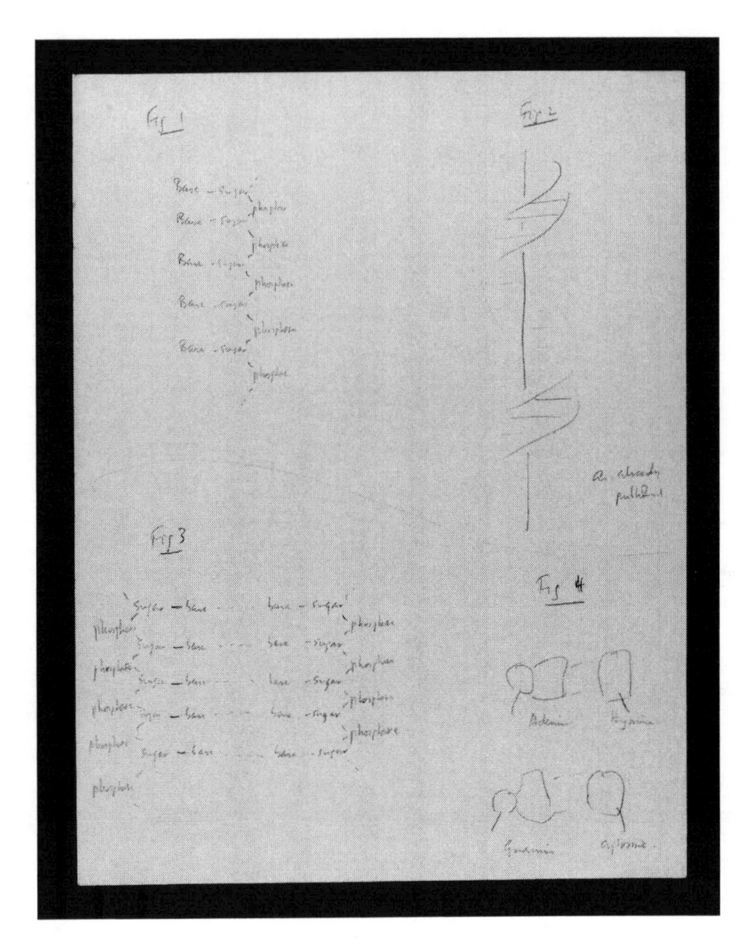

Figure 2.2[2].

victorious heroes of modern biology. For them it conjures the memory of Rosalind Franklin, the physical chemist and molecular biologist who did much of the painstaking X-ray crystallography work from which the two Nobel laureates made their deductions. Many female scientists have strong views that Rosalind Franklin was shabbily treated by the male-dominated scientific community of the post World War II era, and cast in an unkind light by James Watson in his bestselling book, *The Double Helix*.[3] This dominates their interpretation of the simple sketch. It does not alter the facts of DNA, but it creates an Affect among some of the Audience which may not have been expected.

When scientists or anyone else tell a story, its interpretation and the meaning that is created in the mind of the Audience is personal. It depends on how the story resonates with what the listener already knows, how it Changes what they know.

> 'At that meeting he was struck for the first time by the endless variety of men's minds, which prevents a truth from ever presenting itself identically to two persons. Even those members who seemed to be on his side understood him in their own way, with limitations and alterations he could not agree to, as what he always wanted most was to convey his thought to others just as he himself understood it.'
>
> Thoughts of Pierre Bezukhov on giving a speech to the freemasons – from Tolstoy's *War and Peace*[4]

Science is an evidence-based means of understanding the World. It relies on rigorous processes to provide a frame for looking at a particular question and observing what happens. When Change is observed in an experiment, as in a story, the interpretation of what it means begins. The Change is not always self-evident. There might be interpretations which generate more than a single idea, or even conflicting ideas. Indeed, many stories accommodate multiple views and emphases, which is why they are so effective in communicating ideas which are both complex and ambiguous.

> 'Keep your enthusiasm but let verification be its constant companion.'
>
> Louis Pasteur

New evidence is assessed for its perceived quality; it is weighed up against what is already understood. If some new evidence is revealed which is unexpected and flies against a commonly held assumption, then it is likely to take some time before it becomes accepted. Opinions have a broad base: they draw on logic, experience, rationale, intuition and emotions. They consider the reliability and credibility of an individual daring to challenge the status quo by presenting Change.

Over time, the Change and Affect of the central narratives in science move to a consensus. It is not always rapid: Change might not be subjective, but there may be many reasons for explaining how and why it came about.

What is Your Purpose?

If you are asked to consider the purpose of your talk, paper, poster or any other form of communication, your list might include:

to inform
to teach or educate
to share knowledge
to show that my work has significance.

All of these are worthy purposes. They are not, however, much use in helping you create meaning for your Audience or giving them a new way of looking at the World. So, is there a way to broaden the frame for looking at purpose? Let us consider purpose through another question: 'Why do I want to inform, educate or share my knowledge?' The list of purposes should increase significantly.

For example:

to persuade
to provoke
to encourage
to impress
to make a proposition
to inspire
to entertain
to provide a vision.
to illustrate a theory
to correct a mistake
to sell oneself
to create interest in oneself
to give an opinion
to make a point
to test an idea

to preserve cultural memory
to get attention
to get funding
to get support
to excite
to warn
to cement group identity
to provide *crackle* **and** *fizz* **to a discourse**

These two lists can apply either to the overall communication or to just some part of it. The British comic actor John Cleese was interviewed on television some years ago and was asked about the technique of Christopher Morahan, a film director he admired and with whom he had made the film *Clockwise*. Cleese recounted how Morahan, when planning how to shoot each scene, used to ask: 'What is the point of this scene?' His plans, according to Cleese, were 'to shoot the scene to illustrate the point'. The bigger point that Cleese was making (or at least my interpretation of it) was that there are many ways of covering a scene in a film: in wide-shot, in close-up, from the point of view of the open door, the dog or the main character. It can be long and lingering or short and snappy. It can draw attention to a detail: an invitation on the mantelpiece, the level of liquid in a bottle, the sound of humming from someone out of sight. The way the scene is put together can, when crafted skilfully, lead you to laugh or cry. It can make you feel curious or excited. Of course, when it is done without due thought and attention, it can be confusing, opaque or just dull.

If your intention is clear – you want to draw attention to a detail which will become important later on, or you want to impress the Audience with your ingenuity, for example – it gives you a focus for thinking about *how* you might present the scene to meet your purpose.

The same question, 'What is the point of this?', can be asked of each section of a presentation, each paragraph, or graphic or text box on a poster. It also applies to each slide.

If, for example, your purpose is to show the change in concentration of a particular protein over the course of 24 hours, your graph of time against concentration should show this clearly and without distractions of other information which might be relevant to you and your work, but is not directly related to the point you are making to an Audience. If your purpose is to show that you collected readings for a whole 24 hours and nothing happened until the last hour, you might want to reveal the readings in three hour blocks. After seven reveals where nothing changes, the final revelation makes an impact.

Harnessing Change Models – Elevating the Particular to Illustrate the General

Some stories can be told in such a way that their meaning resonates with something much larger, some universal truth about life. To do this you need to begin to think of the Change at the heart of your communication as an example of something bigger than itself. Ask yourself whether you can describe the Change in a research project by using a Change model with which people are already familiar – a narrative model.

The behaviour of a protein in a biochemical pathway might, for example, illustrate a more familiar system, that of enmity to collaboration. The discovery of the importance of a molecule, which previously had been ignored, might have parallels with the Change at the heart of the story of the ugly duckling. The Changes in a cell might, with minor adjustments, also be an account of what has happened to the laboratory over the past few months. The skill is to describe the Change at the heart of the science communication in a way which resonates with what the Audience already knows. Metaphors and analogies are one way to do this.

Stories themselves are, and always have been, powerful metaphors for life's experiences. They are one of the principal means by which all of us, at all ages, learn to make sense of the world, its complexity and ambiguities. Narrative, or the means of telling stories, is an effective way to provide meaning; it requires, however, that Change is presented in a way which the Audience can grasp. Harnessing analogies and metaphors gives the Audience a handle.

It is up to you to choose an appropriate metaphor and to be clear about where it breaks down in describing your research. The *World* chapter covers analogies and metaphors in more detail.

Change Versus Affect – Affect Will Win

A researcher working in the area of synthetic biology was telling a group of colleagues how it would soon be possible to manufacture synthetic meat on a large scale. The only problem, he reported, was that people said they would not want to eat it. This, he announced, was completely illogical. After all, synthetic meat would eradicate the need to kill animals for food, which would surely meet with most people's approval. The synthetic material could also be made to have the same taste and texture of meat. So why would people not want to eat it?

Change is only accepted when the Affect it creates in the hearts and minds of the Audience is preferable to the status quo. Emotions are more powerful a persuader than facts. Communications do not make an impact with logic or rationality; they make an impact with Affect. This is something which needs to be considered in the planning of any communication.

To be clear, I am not suggesting that researchers should aim to create an emotional Affect not directly related to the Change they are communicating. What I am suggesting is that researchers use all the evidence of Change at their disposal to create a narrative which ensures the research findings make the impact with the Audience that they deserve.

This reinforces the need, discussed at the start of the chapter, to be clear about the purpose of your communication. In other words, once your Audience has read your abstract, looked at your poster or heard your presentation, then:

What do you want your Audience to think?
What do you want your Audience to feel?
What do you want your Audience to do?

The Change and Affect is the headline summary of the communication. The Audience understands something new and feels different, but the Change and Affect give us no information about the details of the Audience's journey. In the following chapters we move on to look at each component of that narrative journey.

We conclude this chapter with some exercises to help you use Change and Affect to define the heart of your communication. The questions can help you shape your communication to meet your intentions, your purposes.

Questions and Ideas to Help Identify Overt Change

1. Outline the key features at the beginning of your story in one column. This is your starting World. In a second column outline the end of your communication. This is your finishing World.
    - What is different about the two Worlds?
    - What has Changed?
    - If nothing has changed, what were you expecting to Change?

2. How did the Change come about?
    - Was it a direct observation?
    - Was it manufactured by circumstances?
    - Was it as a result of your actions?

3. What Changed for you?
    - What did you learn on the journey?
    - Why is it significant to you?
    - What difference will it make to you?

Questions and Ideas to Help Identify Similar Change Models

The following can be useful for helping spark ideas for giving you a narrative frame with which you might design a poster or a presentation.

This is a general list of Change models. Are there any which you could harness to your purpose?

Starting World	⟶ ⟵	**Finishing World**
Life		Death
Sickness		Health
Chaos		Order
Playing safe		Risk taking
Demolishing		Building
Revered		Reviled
Inefficient		Efficient
Rags		Riches
Hindered		Helping
Low status		High Status
Powerless		Powerful
Weak		Strong
Bad		Good
Success	⟶	Failure
Complexity		Simplicity
Low esteem	⟵	High esteem
Loyalty		Betrayal
Freedom		Slavery
Cowardice		Courage
Isolation		Community
Ordinary		Extraordinary
Fair		Unfair
Arrogant		Humble
Disaster		Triumph
Enmity		Cooperation
Ugly		Beautiful
Injustice		Justice
Powerlesssness		Control
Feckless		Integrity
Ignorant		Knowing

1. Looking through your list of Changes so far, are there any Change models which immediately strike you as relevant to your narrative? Clearly the choice will also depend on the kind of journey you took from the starting World to the finishing World.
   - Are there any similarities with Changes found in other parts of your area of research in the past?
   - Are there resonances with Changes in any area of science, engineering or technology?
   - Are the Changes in your list also seen elsewhere in a broader cultural context: in popular fiction, in the news, in politics?

2. If you have tried these and no ideas are forthcoming suspend judgement and play with some creative techniques.

   Try forcing a connection between arbitrary Change models (High status/Low status for example). Attempt to define your overt Change as a shift from high to low status, or powerlessness to power.

   Look at your Change from a completely different perspective. Perhaps from the point of view of an object, protein, patient or clinician within the narrative.

   - Do these techniques generate any ideas for constructing the overarching Change in a different way which might appeal to the Audience but still allow you to make clear what you have found or not found?

Questions and Ideas to Help Identify Affect

The Affect is the feeling you want to leave your Audience with. If it is powerful, it might spur them to action: to read your paper, to discuss with colleagues what they have heard. At the very least, you hope that your Audience will think about what they have read or heard, and be able to recall the Change at the heart of the communication.

1. Does the Change you have found (or not found) give you overall a positive feeling?
   - What does it make you feel?
   - Why is that? Can you identify the significant factors in order to make sure that these are brought out in your communication?

2. Could the Change be interpreted by others in a completely different way?
   - What might be its Affect on others?

3. As part of a creative exercise to explore the limits of possible ways to tell the story: how might you tell the story of the Change if you wanted to create a different Affect? For example: inspire instead of impress; provoke instead of reassure; amuse instead of worry.

3

Lure: an enticement, usually with a bait, offering the promise of a reward

Your post has arrived. You notice that you have two letters. One is clearly from a political party. There is its brightly coloured logo on the front and 'Vote for Emma Brown' emblazoned diagonally across the top left-hand corner. You toss this straight into the paper-recycling bin. The other letter, addressed to you, is written in a hand you do not immediately recognise. You try to think who might be writing to you and why. You look closely at the post mark wondering where it was sent from. Opening it reveals a letter and a leaflet; you discover that it is from another political party. You may at this stage still throw the material away, however the handwritten address on the front of the letter had done its job. It had *lured* you into opening the envelope.

If you have a captive Audience and people are keen to listen to or read your message, a Lure may not be necessary. Where people have choices and competing demands on their time, however, then sparking their curiosity, making them pause to consider whether to give you and your communication more attention, is the job of your Lure.

In this chapter we shall be looking at examples of Lures which professional communicators use to grab the attention of their readers, buyers, donors, viewers or listeners. How do journalists attract readers to their copy? What techniques do marketing companies deploy to attract buyers to different propositions? How do charities position themselves to encourage people to donate?

While the examples are unlikely to be transferable wholesale and unedited to specialists' communications, the illustrations should stimulate ideas by presenting a range of approaches to rouse the curiosity of sometimes difficult-to-attract Audiences.

At the end of the chapter there are a series of questions to apply to your particular communication challenge, whether it is designing a poster, constructing a talk or even thinking of a suitable title for a paper. The individual answers will hopefully spark further ideas to help you construct and refine an effective Lure.

The Lure is an opportunity to tempt people to give their attention to the main communication event. It is the only narrative ingredient which must appear either before the event itself or at its very beginning. In the broadcast and film worlds it is the equivalent of the trailer.

Why do Scientists Need a Lure?

A glance down the agenda of most conferences attended by specialists reveals a list of titles of talks, posters and sessions which have been prepared with an Audience of 'colleagues' in mind. That is they are generally statements which summarise a niche scientific area. If they contain scientific jargon, it is a shorthand with which – it is assumed – most of the Audience will be familiar.

Scientists who make a conscious effort to produce a Lure – an eye-catching poster or a title which manages to keep jargon to a minimum – are likely to attract more people to come and see or hear their content. Below are the titles of competing presentations which ran in parallel at a leadership conference[1] attended by professional coaches.

No prizes for guessing which of the sessions had the biggest Audience!

Coaching gender diverse groups for leadership success.

Dr Savita Kumra

Coaching for excellence in strategic leadership.

Christine K Champion

Leadership coaching in the transpersonal dimension.

Dr Tatiana Bachkirova

Confessions of a Leadership Coach.

Jon Davidge

The first three titles, at face value, are quite narrowly specific: gender diverse groups, transpersonal dimensions, strategic leadership. They mean something to the speakers and some of their Audience, but they are also jargon; and if you don't have any more information about the track record of the speakers, it is difficult to choose between them. For a casual attendee, none are appealing.

Confessions of a Leadership Coach could be the title of a *Carry On* film. Although it appears unspecific – it could be about anything – the word 'confession' promises some embarrassing or guilty disclosures. *Confessions of a Leadership Coach* had the highest attendance and additional seating needed to be brought in. Professional coaches, like everyone else, are not immune to being curious about the weaknesses of others. Importantly, the speaker addressed his title in his opening words. He admitted that he had chosen it with the express purpose of attracting a large Audience, but went on to talk about some of the dilemmas which professional coaches are faced with in their work. The Audience of specialists and interested others was not disappointed. This Lure met all four criteria for maximum effectiveness.

LURES must be Audience specific
LURES must be noticed
LURES must be relevant
LURES must have *crackle* and *fizz*

Lures Must be Audience Specific

'MAYDAY, MAYDAY, MAYDAY' heard over a crackly radio is a Lure which will instantly alert sailors to give their full and undivided attention to what they hear next. This is an example of a Lure which will, without fail, capture the attention of a specific Audience (sailors, pilots) because professional training has programmed that Audience to respond in a particular way. Church bells chiming on Sunday mornings, umbrellas held high by tour guides and ice-cream vans blaring jolly tunes are all Lures which work because their intended Audience is primed to respond to them. It is the same reflex response which was first demonstrated by the physiologist Ivan Pavlov when his dog produced saliva in response to a bell, having previously associated the bell with feeding time. The Lure can act as the most powerful of heralds to an Audience with strong, preformed associations.

Unfortunately, for most experts communicating specialist knowledge, there will be many occasions when your Audience is not so primed. The Lure will have to work much harder to draw people to your content.

The Lure can be an advertisement or pre-publicity for your communication, or it can be your opening words, first slide or a demonstration. For a poster presentation the Lure is the title or picture or layout which draws the Audience to give it further attention.

Pre-publicity

There are many ways to signal to an Audience that you have something interesting to say:

> posters, blogs, texts, flyers, a synopsis in an agenda, a personal recommendation – particularly from a high profile individual, a press release, the offer of a free lunch, a catchy title or headline.

Designing a Lure

Think back to when you were ten years of age and choosing books to borrow from the library. What criteria did you use to make your specific choices? Previous experience with an author? A bright, upbeat colour picture on the cover? Print size of not less than 14 point? Not excessively long? Someone's recommendation? Its appearance on the returns trolley? Its popularity – deduced by the number of date stamps on the inside cover? Its title?

How a child chooses books may at first glance seem hardly relevant to giving a scientific talk – unless one is planning to give a talk to ten year olds. I chose this example because with some thought we are all able to think back to what it was like to be ten. Being able to see the world from different perspectives is a key creative skill, hugely useful in all communication because it is essential for connecting with an Audience.

Just as fishermen will select a different kind of lure to draw a particular type of fish to their hooks, based on their knowledge of fish behaviour, so too will storytellers tailor their Lures to capture the attention of a specific Audience.

At the end of the chapter there are questions to ask yourself about your Audience to help stimulate ideas for Lures. First, an example from a professional communicator – not an advertising planner as in the last chapter, but a tabloid newspaper editor. You may think that tabloid newspapers are so far removed from scientific papers and posters that their approach bears no relevance to scientific communications, yet tabloid newspaper editors understand their Audience. They recognise those features, words, pictures which will attract readers to buy their papers. Journalists who write articles for newspapers rarely write the headline. These are written by specialist writers who have a knack for using words sparsely, both to convey meaning and arouse curiosity. Nowhere is this demonstrated more clearly than on the front pages: arresting photographs and headlines in large font make up a significant proportion of the area. Each of the national tabloid dailies can be read at a hundred paces, from across a garage forecourt or a railway platform.

DAILY STAR

THE No. 1 PAPER FOR NEWS

WEDNESDAY, MAY 25, 1994

25p (26p Cls)

OOH, AAH! DAILY STAR WIN A ZIPPY YAMAHA

See Page 13

SIN-ITTA!

Her love for Liz Hurley's ex-hunk

EXCLUSIVE

See Page 15

MISS MATCH: Gazza

Love split Gazza on a benda

IAN TRUEMAN

KILLER BUG

PETER BOND

THIS man is a walking miracle.

Ken Kruck had 400 stitches and six horrifying operations after falling victim to the flesh-eating sci-fi bug.

But his astonishing survival against all odds was set yesterday against a young woman who died from the nightmare disease in a London hospital.

She was the SEVENTH person to die this year from the killer bacteria known as necrotising facitis. Yesterday the Government was urged in the Commons to take every possible step to calm public fears.

Labour Health spokesman David Blunkett stormed: "We want to know exactly what is being done."

But doctors admit they are baffled.

Picture: DOUG DOIG

EXCLUSIVE

Sci-fi horror: Pages 4, 5 and 8

ATE MY FACE

Figure 3.1.

Figure 3.1 shows my all-time favourite: 'Killer Bug Ate My Face', a headline in 1994 when necrotising fasciitis was in the news. I defy anyone seeing this front page not to be drawn to read the whole article. The *Daily Star*[2] found Niko Kraket when he was suffering from a necrotising fasciitis infection and gained his consent to have his photograph on the front page. Beside him there is also a picture of a micro-organism – a rare event for the front page of a red top.

Your first reaction may be that such graphic headlines and gruesome pictures are neither desirable nor suitable for your communications, but ask yourself what will catch the attention of your desired Audience? Can you link your content to something you know will arouse their interest? Are you able to encapsulate or perhaps hint at the essence of your content with a picture or with some carefully chosen words?

Lures Must be Noticed

Lures fail if they require concentration or effort to make their impact. Lures are noticed because they stand out. They do so either because the Audience has been primed to notice them, as in the MAYDAY example, or because they are novel, surprising or bold.

Novelty, surprise and *boldness* may seem blindingly obvious attributes for a Lure, yet they are often lacking in science communication even when the Audience are the public or non-specialists. Life is made easier by conforming to the expectations of peers but it does not entice Audiences. In fact, the reverse is true. The aim of the Lure is to challenge the expectations of the Audience, whoever they are, by doing something unexpected.

Novelty

Keep very still and listen.

> What can you hear? What else?
> How many of those different sounds are always there?

How often are you aware of them?
What makes you aware of them?

If after reading this sentence you shut your eyes what new sounds do you hear?

It is a matter of survival that human beings and animals notice newness in case it signals a brutal end to our activity – or life. By the same rule we all become quickly accustomed to routine events which can soon become ignored as unimportant or unremarkable. Novelty is short lived. This is why advertising agencies and entrepreneurs are kept in business. The challenge is to continue to find new ways to capture people's attention. Nowhere is this more important than for people whose job it is to raise money for charity.

Figure 3.2[3] is a photograph which came in an envelope addressed to my husband.

On the back of the picture of the boy is a message.

Figure 3.2

I took this picture of Gach Chol myself. The day he walked out of the MSF project, under his own steam, was marvellous.

My husband asked me if we knew anyone who might have sent it. We looked in the rest of the envelope and of course found the details of how to give money to the charity: Médecins Sans Frontières. The Lure had done its job. It had novelty. It will, however, be unlikely to work again in the same way. If you have developed a successful Lure, others will copy your idea and soon the Lure will become less effective. You will need a new one.

> 'Novelty, has charms that our minds can hardly withstand.'
> Attributed to William Makepeace Thackeray

Surprise

We notice newness and we are surprised by the unexpected. They grab our attention. So what is the most unexpected fact or thing that you can surprise your Audience with in your Lure? It can be an unexpected question, a surprising statistic or, if it is a talk or presentation, a prop; something physical which the Audience is curious about.

An interviewee applying for a researcher job in BBC Science came in with a briefcase and a thermos flask which he very carefully put down on the floor beside his chair. He did not refer to the flask immediately, but the selection panel were all intrigued to know whether there was something other than coffee inside. When asked about his ideas for popularising science he opened his brief case, put on asbestos gloves and took up the flask which we discovered contained liquid nitrogen. He gave the panel a simple demonstration and explanation of the effect of rapid freezing on a grape. It was an effective way of capturing attention.

A surgeon talking about a procedure he had pioneered had a different 'prop' which could also be described as a Lure. He appeared on stage with one of his success stories – a patient. Something real and unexpected, presented to the Audience can be sufficient to grab their attention.

> 'Two years after my mother died, my father fell in love with a glamorous blonde Ukrainian divorcée. He was eighty-four and she was thirty-six. She exploded into our lives like a fluffy pink grenade, churning up the murky water, bringing to the surface a sludge of sloughed-off memories, giving the family ghosts a kick up the backside. It all started with a phone call.'
>
> Marina Lewycka[4]

This is the copy on the back cover of the novel *A Short History of Tractors in Ukrainian*. It is also its first paragraph. The title is unusual for a novel and the copy is intriguing and full of colour. Together they make for a compelling Lure, and one that will attract readers to take a closer look.

Boldness

It is not always enough for a Lure to be novel and unexpected. It must also stand out like a beacon. Its job is to signal above all the random noise which surrounds the Audience: 'Hey, I am over here! Come hither.' A Lure must be bold.

Boldness personally demonstrated by individuals can also act as an effective Lure, as shown by the example below.

A colleague of mine, Lorraine, wanted to publicise a brand-new BBC television series called *Fat Nation,* which she was promoting ahead of its transmission. When the allotted time came for her presentation to several hundred television professionals, the astonished Audience noticed a slight kerfuffle in the wings and watched as she staggered onto centre stage. Lorraine is a little over five foot tall and dresses with panache. On this occasion she was wearing a fat suit beneath an ill-fitting skirt and blouse and looked quite grotesque.

Later there was much discussion about the suitability of her stunt, whether it was in good taste or not, but at the time, she had the full attention of every one of her colleagues. Her introduction to the series was remembered by all who witnessed the Lure and possibly for longer than the series itself. By nature, Lorraine is an extrovert. It can be much more difficult for introverts who do not enjoy the limelight. Nevertheless it took boldness and courage to pull off a stunt like that.

> 'Behold the Turtle. It makes progress only when it sticks its neck out.'
>
> James Bryant Conant, Chemist[5]

Research scientists, particularly junior scientists, often express concern that if they were to take a bold approach in designing their Lure or in introducing their talk, they would be knocked down. It is not done to show off and draw attention to oneself in research. There is a hierarchy in science and so, although it might not be acceptable for junior scientists to produce bold Lures, senior scientists who are good communicators invariably do. These scientists are more practised and relaxed at giving presentations; they have developed the confidence to be bold. Learning to be brave is part of the process of becoming a better designer of effective Lures.

There may be occasional situations when it would be foolish to take a risk with a bold Lure. A more informal event, however, or where you are presenting material to your immediate colleagues, are both opportunities to try out something new. It may not always work but if it does, you will be able to add the idea to your repertoire for bigger occasions. As you get more practice you become more experienced and more confident in front of your peers and senior colleagues. However, if you are planning to give a talk to the general public, they have no such qualms about scientists drawing attention to themselves with a striking Lure. They will want to think that they are going to be entertained by hearing something new and important and the initial Lure must be designed with them in mind and must be bold enough to catch their attention. It must, of course, also be relevant to the content of the presentation or talk.

Lures Must be Relevant

There is absolutely no point in having an alluring piece of bait to get people to want to listen or read your content only to find that the Lure has no relevance to the content. No-one likes being conned. When I was editor of a science series I received a proposal from a scientist with the headline, 'Great Plague: London in danger NOW'. After reading the short proposal I met with the scientist and it became clear that his proposal was so far removed from any likelihood of happening, it bordered on fabrication. There was, and is, an interesting story about rats in the cities of the UK but to suggest that the Plague was about to return was very misleading.

For specialists, the most consistent opportunities for constructing a Lure which will stand out, be noticed by the Audience and also be relevant, will be the title.

Titles

In the first 30 years of television, when there were only a few terrestrial channels, many of the titles of the major series were fairly bland: *Panorama*, *Horizon*, *Everyman*, *Grandstand*, *Omnibus*. They conveyed very little of the individual programme content. This did not matter very much as there was no competition.

With the glut of channels we have today, however, there are a bewildering number of choices to be made about what to watch, and many alternative digital options for entertainment. Several years ago, British programme makers followed the lead of their American counterparts, who were used to a more crowded marketplace. They too became more inventive with their titles to tempt the Audience to tune in: *Law and Order*, *World's Most Embarrassing Bodies*, *How Clean is your House?*, *Bang Goes the Theory*, *Wonders of the Universe*, *Shock and Awe*. All of them let the Audience know what the programmes are about. They are concise or catchy, and relevant.

Almost every new title for content which appears as a book, a film release, a television programme has been agonised over by people who know the potential power which a title has to Lure an Audience.

When titles are relevant they can also help give a focus to the whole communication. The tone of a communication and its content can sometimes sprawl and lose focus in the preparation and planning. By spending time working on a title which is relevant, it can work for you, saving you time later on.

An example is a series which I edited at the BBC. The idea was to look at the history of people from earlier ages through the excavation and investigation of skeletons recovered in archaeological digs. The hope was that we would find different skeletons from the Bronze Age through to the Vikings and be able to look at the history of these different periods through information we could extract from the forensic examination of the bones and the contexts of the burials. The title of this idea was *Meet the Ancestors*. The commissioners, who control the budgets and therefore decide whether a programme will be made, liked the title so much that they pretty much funded the programme on the title alone.

The title soon started to give the production team a focus. In the series we would 'meet ancestors'. That implied seeing what they looked like and finding out as much about them as we could. The focal point for the programme changed very early on during filming from being about the Bronze Age, Iron Age and Roman era to focusing on stories of individuals from the past. You may think that this was an obvious way forward, but believe me it was not at all clearly defined in our minds when we first envisaged the programme. We were wedded to telling stories of a period. By allowing the title to guide us we had to abandon some of the early skeletons we had found, which were interesting in the diseases and conditions they demonstrated, but which had no skulls and were therefore not 'a person we could meet'.

Titles not only need to be relevant, they can also be a guide for deciding on the relevance of the content. Does it fulfil the promise of the title?

Here are the titles of lectures[6] given by young engineers and scientists. All were chosen by the British Science Association as having demonstrated outstanding skills in communicating to non-specialist Audiences.

Can marketers control your mind? Facts and fiction about Neuromarketing and the 'Buy button' in your brain.

Prof. Nick Lee

Hit me with your rhythm schtick: the connection between music, movement and the brain.

Dr Jessica Grahn

Let George do it: are we overautomating our lives?

Dr Mark Young

Watering thirsty cities.

Dr Sarah Bell

The Big Bang dilemma.

Suzie Sheehy

This is a lecture title which someone sent to me when I was preparing material for a lecture on the public understanding of science:

Chimeric methylphosphonodiester/phosphodiester antisense oligeodeoxynucleotide analogue: Enhanced Activity and Specificity at Directing RNase H.

Richard Giles, University of Liverpool.

- ✓ Lures must be Audience specific: this is directed at a particular Audience, albeit a very narrow one.
- ✓ Lures must stand out: the extraordinary length and unusualness draws attention to itself.
- ✓ Lures must be relevant: not only is it relevant; it gives no opportunities for any misunderstanding.
- × Lures must have *crackle and fizz*.

The problem for me with this title is that it lacks *crackle and fizz*. (Having said that, I might be tempted to go to this out of pure curiosity to see the speaker who devised this title, but I would sit near the door and be ready to creep out once I had satisfied my interest.)

So what is *crackle and fizz*? It is the extra ingredient, often ambiguous, which gives a serviceable Lure added appeal. A Lure with *crackle and fizz* will not only stand out, be relevant to the content and targeted to an Audience, it has another quality. It generates a whiff of pleasurable anticipation, engages the Audience's imaginations as well as their intellect. It has a cleverness to it which pleases the Audience and leaves them looking forward to engaging further with the content.

To use an analogy, it is the difference between a menu which lists:

Cod with chips and peas

and one which lists:

Fresh line-caught Atlantic cod with crunchy thin-cut French fried chips and Moroccan minted peas

The first menu has clarity, the second vividly engages the imagination.

When I originally conceived the idea for this book I came up with many different titles, most designed to include the important search keywords: research scientists, communication skills, persuasive skills, presentation techniques, pitching skills. I played with dozens of different permutations and tried them out on people. Their expressions confirmed what my instincts had already told me: they all sounded rather dull. I wanted this book to be lively and readable. I wanted the title to express or at least evoke an emotion, which is critical to effective communication. *Crackle* and *fizz* are what good presentations should have. Lots of sparkling ideas and thought provoking stories. I checked *Crackle and Fizz: Essential Communication and Pitching Skills for Scientists* against the four criteria for a successful Lure:

✓ Lures must be Audience specific: scientists looking for help with constructing presentations to the general public will not make the association between *crackle and fizz* and communication. However the sub-title has relevant words for search engines.

✓ Lures must stand out.
✓ Lures must be relevant.
✓ Lures must have *crackle and fizz*. This is a highly desirable quality – but only if it is Audience specific, it stands out and is relevant.

Designing titles which will attract and draw in an Audience is not always easy. It takes time, but there are a number of creative techniques which can help focus your thoughts and give you ideas. These are at the end of the chapter.

Lures Must Have *Crackle and Fizz*

André Geim and Kostya Novoselov won the 2010 Nobel Prize for Physics for isolating single layers of carbon – graphene. Earlier in his career, André Geim had demonstrated that all materials have some magnetism. He showed that if you put various materials in a strong magnetic field they float. He demonstrated the phenomenon with droplets of water, strawberries, grapes, insects and a frog. He was demonstrating the phenomena of diamagnetic levitation. His choice of levitating a frog – which was unharmed during the demonstration – captured the imagination of Audiences. It was, he said, the 'wow factor' which you need to attract attention, to Lure your Audience.

> 'We were amazed to find out that 90% of our colleagues did not believe that we were not joking that water can levitate. It became obvious to us that it was important to make scientists (as well as non-scientists) aware of the phenomenon. We levitated a live frog and other not-very-scientific objects because of their obvious appeal to a broader audience and in the hope that researchers from various disciplines, not only physicists, would never ever forget this often neglected force and the opportunities it offers.
> In addition, the frog picture will probably help students studying magnetism to get less easily bored.'
>
> André Geim[7]

And Finally

Personally, my best story of designing and delivering a Lure was one where I felt I was taking a big risk.

I had just been given a new job of introducing an innovation process into the BBC and I was called to a meeting of senior managers to tell them about my role. They were a formidable group and I needed to have them on my side if I was to be successful in my task.

There were 20 or so people in the room, and many had only half an ear on the proceedings as they were also engaged with e-mails and their mobile phones. The chairperson introduced me and a few people looked up briefly. While they did, I held up a banana and I held it in the air until everyone else was looking up too. 'Do you know', I began, 'that most people peel a banana from the stalk end, but monkeys always peel them from the other end?' I had the full attention of every person in the room. I used the banana and the question as a Lure to get their attention but I also used what I said next to take them straight into the World of my story – which was the organisation we all worked for. I drew attention to one part of that World, and one which they recognised; a place where, too often, people did things in a particular way because that was the way it had always been done. I then outlined my project – to help people to be more creative about their work. I ended by telling them that people peel bananas from the stalk end because that is the way they have always done it and they *think* it is easier, but monkeys peel from the other end because that is the end bananas ripen from – and so it *is* easier.

The Lure was made Audience specific. It was novel, surprising and bold. It was made relevant to the content. Put simply, it aroused the curiosity of the Audience so that giving their full undivided attention to the message did not seem in any way a chore. I connected with the Audience and immediately launched into the World of my narrative – the subject of the next chapter.

Questions and Ideas to Help Stimulate Ideas for a Lure

Below is a series of questions to help you consider suitable Lures. Not all of these will be relevant to your individual communication, but write down as many answers as you can to the questions before going back to judge their quality or relevance. After this you can hone them to best suit your communication.

Remember, in planning be as bold as your imagination allows. It is always easier to tone down something which is plainly over-the-top than to inject life into something which is rather dull.

Your aim is to be as bold as you dare for every Audience.

Designing a title

1. Whose attention do you want for your piece of communication, your presentation, your story?

(Refer back to the questions at the end of the *Audience* chapter)

2. Write down individual sentences; each attempting to summarise why you are communicating this specialist knowledge.

Hints:

- What new evidence are you presenting?
- What is or was your project designed to do?
- Is there a particular question in your topic area that the Audience is likely to be curious about?

3. How would the following paraphrase each of your sentences?
   - Your supervisor or head of department.
   - A ten-year-old child.
   - A character (pick any you feel you know well). They can be from public life, literature or even a soap opera.
   - *Nature* or popular science magazines.
   - A tabloid newspaper headline writer.

4. List some interesting random places, people and objects very familiar to your Audience. (If you are at a conference, include those hotel facilities or particular arrangements which have caused discussion among the delegates.)

Take each of the sentences you have written down in response to points two and three and force a connection between them and the random items in your final list.

- Can you adapt the sentences to describe some activity, relationship or mechanism related to those well-known places, people or objects?
- Are there any well-known (to your Audience) phrases which spring to mind which you could adapt for a title or an introduction?

If you find this difficult, try the following exercise to get your brain sparking with ideas. Force your imagination to find three things which two random pictures or objects have in common. You can find pictures in any magazine. Take a single picture and force three connections between it and one of the following: a frog, a lump of coal, a kite.

5. List any relevant facts or statistics associated with your topic (in its widest view).
   - Which of these will most surprise or impress your Audience?
   - Can you form a question or a context for any of these facts which could make an intriguing title?

The exercise at the end of the *Change and Affect* chapter may also give ideas for an effective title. It is often a good idea to work through a first rough draft of the communication before designing or committing to a title.

Pre-publicity

This can be a physical or electronic display: agenda items, synopses, posters, flyers, blogs, texts or Twitter. It can also be through word of mouth.

1. How will your Audience discover that you have some communication to impart? What could you do to improve their chances of discovery?

Hints:

- Are there people who could be recruited, asked or persuaded to recommend you and your communication?
- Is there a striking picture which is directly relevant, or could act as a metaphor for your message which you could use in a physical display?
- Are there lists of facts, statistics from question five in the titles section which could form a montage in the pre-publicity?

2. If the Lure has a physical presence, where will it be placed?

Hints:

- Does it have competition? If so, what is its competition?
- How can you make yours different so that it stands out?

3. Are there any ideas from the exercise on titles which you could adapt for your pre-publicity?

Start of the communication

1. Is there a short anecdote or story you can tell which captures the mood, tone or context of your specialist communication? (There is more on this in the *World* section.)
2. Is your summary of the Change and Affect an effective Lure?

3. What are the highlights of your communication? What is the biggest view of your subject, and the narrowest focus which your communication will encompass. Can you link them to provide a descriptive scope for the journey which your narrative will take?
4. What are the key questions which the Audience have on your topic? List them and give hints to how you might provide answers?
5. Is there some physical object (or description of one, if your communication is written): a model, a demonstration, a picture or a sound recording which you can bring to a presentation or describe in words which you can make relevant to your content?

4

World: the contextual space in which your narrative will unfold

'MAYDAY, MAYDAY, MAYDAY.

This is Yacht Fizz, Yacht Fizz, Yacht Fizz.

MAYDAY.

Our position is 51°30'26"N : 0°06'22"W. We have a hole in our bow and are sinking.

We are six people on board, one unconscious. We are taking to the lifeboat with portable VHS.

This is Yacht Fizz, OVER.'

The 'MAYDAY' distress signal, which I referred to briefly in the previous chapter, is formatted to give potential rescuers essential information briefly and with the utmost clarity. In essence it gives listeners a snapshot of the World of the sinking yacht. First, it includes the position of the vessel; second, the nature of the distress; and third, the numbers involved – the scale of the distress. Potential rescuers with a geo-positioning satellite device would know exactly where the drama is located. They would need no other information.

In this fictional 'MAYDAY' call, however, the sinking yacht is not at sea; rather it is being watched by hundreds of people lining the banks of the river Thames in London. The more effective communication strategy for the skipper in this location, therefore, is not the one she was trained to use to attract attention at sea. Here, she simply needs to yell 'help' to raise the alarm.

My intention with this story is to illustrate that while there are times when it is entirely appropriate for experts to communicate in various types of shorthand that excludes all others, there are equally occasions when it is not.

Science communication, like the 'MAYDAY' call, is also strictly formatted. Theses and science papers need to be written in a particular way so they comply with the expectations of fellow experts: examiners, reviewers and readers. It is rarely appropriate, however, to use the same approach in communicating with a wider Audience.

Scientists' narratives take place in physical and mental Worlds which bear little resemblance to the everyday World of the majority. If outsiders, including junior colleagues, researchers with different specialist knowledge, the public and funders, are to understand what you do, then they will first need some idea of the kind of scientific World you inhabit. A different approach to communication is needed.

The more strange and unfamiliar your territory, the more resourceful you will need to be to take your Audience with you on your narrative journey.

Finding ways to cross the communication chasm between the ordinary World and specialists' World is the subject of this chapter.

What do you make of the text in the box? Can you guess what the procedure refers to? What is going through your mind as you read it?

> 'The procedure is actually quite simple. First you arrange the items into different groups. Of course one pile may be quite sufficient depending on how much there is to do. If you have to go somewhere else due to lack of facilities that is the next step; otherwise you are pretty well set. It is important not to overdo things. That is, it is better to do too few things at once than too many. In the short run this may not seem very important but complications can easily arise. A mistake can be expensive as well. At first, the whole procedure will seem complicated. Soon, however, it will
>
> (*Continued*)

(*Continued*)

become just another facet of life. It is difficult to foresee any end to the necessity for this task in the immediate future, but then one can never tell. After the procedure is completed one arranges the materials into different groups again. Then, they can be put in to their appropriate places. Eventually they will be used once more and the whole cycle will then have to be repeated. However this is part of life.'[1]

Are you thinking this is some kind of assay system? A way to separate chemicals from one another? It is clearly describing something practical in great detail, but without a sense of the context it is confusing and difficult to fathom. As a reader you feel disoriented.

If you are given the word 'laundry', the chances are you will read the text in a completely different light. The content now has a context and is instantly comprehensible, despite being somewhat convoluted. When I have shown this to researchers in my workshops, some of the men have been known to ask, in all seriousness, why it is necessary to 'arrange the items in different groups'. Even when some details are not understood, the narrative – such as it is – can be followed because of some familiarity with the World under discussion.

This chapter will look at various ways to set the scene for your communication. We shall explore how to connect the science World with the ordinary World using metaphors and analogies which, if carefully chosen, can give people some sense of otherwise unfathomable phenomena. Finally we take another look at risk, a World which, without training – and even with it – poses particular challenges for communicators.

At the end of the chapter there are exercises and questions to help you find a number of ways to frame your specialist World so that Audiences can understand the context of your narrative communication.

Set the Scene

This next story is an analogy for communication. In particular, it concerns how to set a scene in order to give the Audience a sense of scale.

An observer watched four men working alongside a trench. Three of them were crouching and one was standing upright. There were several metres between them. Beside each was a square wooden board on small metal wheels which was loaded with a neat array of red clay bricks. On the ground near to each brick trolley was a bucket of cement.

The observer went to the first man and asked him what he was doing. The man held up a brick and expertly shaped a dollop of cement on one side: 'laying bricks' came the reply.

He walked further on, towards the second man, and asked the same question. The man paused, tapped his trowel on the wall and said, 'I am building a wall'.

When the third man was asked, he looked up and into the distance and thought for a moment. Then he turned to the observer. 'I am building a cathedral', he said. The fourth man had a notebook wedged under his arm and was scrutinising sheets covered with drawings and calculations. His name was Christopher Wren. He told the observer he was reconstructing an entire city.

You may be familiar with this story – or some form of it. I find it a useful analogy, not only for how to introduce a scene, but also for thinking about how science is done. As a collective, researchers are contributing to our understanding of the physical and mental Worlds – the equivalent of building the city. There are hundreds of different kinds of scientists all working on different parts of the fabric of the city and its landmarks and spaces. These represent the different scientific disciplines and sub-specialities. Individually, however, research scientists see themselves as working on the 'bricks' of science, and laying foundations for others to build upon.

Analogy – comparison between two things.
e.g. *The immune system responds like a battle-ready army.*

Metaphors – word pictures.
e.g. *His thesis crackled and fizzed with ideas.*

Simile – a figure of speech which compares one thing to another.
e.g. *Her analysis was as sharp as a razor.*

As an analogy for how to view communication, it reminds us of the need to give an Audience a sense of the World in more than one dimension. In my craft as a television producer, we had four basic frames for doing this. The *wide-shot*, which showed the whole scene, the *mid-shot*, which focused on the main object or person, and the *close-up* and the *big close-up*, which put a small area of interest in full view. When film makers want to introduce or create a new World, they do so by using a variety of different shots in different sizes. For example, think of how the World of New York is often conjured in films: Manhattan in aerial wide-shot, or its skyline, followed by close-ups and big close-ups of billboards, street signs, yellow taxi cabs and other well-known features of the city. It is a shorthand way of transporting the Audience to a place. Adults watching a programme are capable of adjusting from looking at a close-up immediately following a wide-shot, without feeling disoriented. Experience of the grammar of television and film has taught them that the two are linked. Children, however, do not automatically make this connection. In programmes where the Audience is young children, the directors are more likely to use the zoom lens. The change between the big view of the whole scene and a close-up detail is achieved through a slow movement. The children have a sense of orientation throughout. They feel part of the scene. When the World being described is new and strange, as it can be with some science communication, we are all like children: we do not automatically connect the big view with the detail. We do not have any experience of it and so we don't know how to make the connection. In science communication the Audience needs multiple views of the World in which the narrative unfolds.

Researchers become experts in their particular area of interest quite quickly – their brick or piece of the wall is their entire focus. They and their colleagues around them know exactly where the wall or brick lies in the big plan; they have a sense of what is around them, but they pay little attention to it. They do not need to. They have defined procedures and ways of working which are routine and understood by all. When visitors come and look at the area and try to understand what is going on, they need some kind of induction to get their bearings and appreciate the detail of the work.

This induction to the narrative World is important for the Audience in any communication. It can be achieved in many different ways: through vivid descriptions of the space and/or the landmarks that allow the Audience to picture the scene in their heads. Skilled communicators can describe scenes moving deftly backwards and forwards from the big picture – the city skyline, to a tiny detail – the initials carved on a single brick of a house in a street. They use metaphors – word pictures – to provide memorable images which feed the imagination of the Audience and make them feel as if they are actually there.

The stranger the World, the more the landmarks and features need to be described. It is quite normal for researchers, who spend their lives working in the lab, to imagine that their surroundings and working environment are not interesting to a non-scientific Audience. In my experience this is not the case. One of the most compelling talks I heard in a workshop was given by a researcher who described his daily routine in fascinating detail. We saw him in his lab, with the sounds of the buzzing lights and the noises of machines. With each new visual sketch, he gave us an insight into his World until we were eagerly curious to know his story, the one he had been preparing us for.

Good communicators gather and cultivate anecdotes and short stories which serve to introduce a subject, give a flavour of the World and tell the Audience something intriguing about it.

When, for the *Royal Institution Christmas Lectures*, our subject was the brain, we struggled for some time about how the lecturer should begin the first lecture. We had planned to start with the Lure of isolated, but interesting, facts and figures about the brain. At the very last minute, I felt we could do better. I asked the lecturer to tell the story she had told at her audition. It was the story set in the USA in 1848; a time when the Americans were laying thousands of miles of railway line across the country. Phineas Gage, a well-respected and capable foreman of a railway construction gang, was preparing the ground for the Rutland and Burlington Railroad in Vermont when he had a terrible accident. His job was to push dynamite down a deep hole with a pole. The pole was more than a metre long and over an inch in diameter at its widest part. On this occasion, 13 September 1848, the charge of dynamite at the bottom of one of these holes ignited early.

It drove the pole through Phineas Gage's head to land 25 metres behind him. As the Boston Post newspaper reported at the time, 'the most singular circumstance connected with this melancholy affair is, that he was alive at two o'clock this afternoon, and in full possession of his reason and free from pain'. He lived another 12 years, was able to work, but his life changed irrevocably because the damage to his brain altered his personality and character. He was no longer the affable fellow he once was. Instead he became fitful, impatient and irritable. His story is the first recorded incident to show something about the relationship between personality and the function of the front part of the brain. It takes the Audience vividly into the World of the brain and demonstrates some of its amazing properties. The story settled the Audience and held them spellbound. Thus, the communication became a two-way process. Instead of pushing facts and figures to the Audience, the story created a pull for information as they became curious about the details, and hungry to know more.

Professional communicators from television and film make trailers for their content. These usually consist of a variety of short clips that rarely tell a story in themselves, but describe the World enough to give a flavour of the mood, the tone, the kind of action and the Characters – that is the World in which the film or programme will take place. *What the Victorians Did for Us* was a BBC series which explored the science and technology of the Victorian era from electric lights to the London sewers, from steam engines to industrial looms. At the start of each programme, before the opening titles, the presenter told a 30-second story relevant to the topic of the programme. It provided the Audience with a sense of the Victorian World, as well as the tone and mood of the programme itself. It gave the Audience a taster, so that they could choose whether or not they wanted to step into the World and watch the programme.

Connect with your Audience's World – *Directly*

Politicians are required to connect with many different constituencies of the public. One British cabinet minister recounted a story of a time he had visited a school to talk to nine-year-olds about politics.

He began to tell them about Mrs Thatcher, the former prime minister. The children did not know Margaret Thatcher; she had retired from politics before they were born. However, they did know Carol Thatcher – Margaret Thatcher's daughter – because she had recently won a popular reality TV contest (*I'm a Celebrity, Get Me Out of Here!*). As the politician talked, he heard a loud whisper in the front row as one of the nine-year-olds turned to her friend and said, 'I think he might be talking about Carol's mum'. The politician later recounted that this interruption was a salutary lesson in being relevant to your Audience, knowing their World.

Beginning in a World your Audience knows can provide creative opportunities for communicating your work or ideas in a completely different way, as this next example demonstrates.

The *9–11 Commission Report*[2] on the attacks on the USA in 2001 ran to 450 pages. It provided readers with a thoroughly detailed account of the terrorist attacks in America – the historical background, and the organisation and management systems in the US government and those between the governmental agencies. It was, and still is, an important document which the authors want to be widely read. It is not, however, the kind of report which parents might buy for their teenage children or one which holidaymakers might pick up at an airport lounge, or even after browsing casually in a bookshop.

However, this does not mean that the report was limited to a very narrow readership. When the content was presented in a different way, it was able to reach a surprising number of people, including those who might have been put off by the sheer weight of the official report, or a perception that it was too technical to be relevant to them.

9/11, The Illustrated 9/11 Commission Report – The Full Story of 9/11 Before, During and After the Attacks[3] by Sid Jacobson and Ernie Colón is 133 pages of cartoons. Cartoons of the kind found in comic books. The only pages without drawings are those with the publisher's details and a foreword written by Thomas H. Keen and Lee H. Hamilton, the chairman and vice-chairman of the *9/11 Commission Report*. They endorsed the graphic adaptation because they knew it could help the lessons learned from their report reach a larger, more diverse readership of US citizens and others across the globe.

FOREWORD TO *9/11: THE ILLUSTRATED 9/11*

(*First two paragraphs*)

COMMISSION REPORT – SID JACOBSON & ERNIE COLÓN[3]

IT WAS THE GOAL OF THE COMMISSION TO TELL THE STORY OF 9/11 IN A WAY THAT THE AMERICAN PEOPLE COULD READ AND UNDERSTAND. WE FELT STRONGLY THAT ONE OF THE MOST IMPORTANT AND TRAGIC EVENTS IN OUR NATION'S HISTORY NEEDED TO BE ACCESSIBLE TO ALL. OUR GOAL IN THE 9/11 COMMISSION REPORT WAS NOT ONLY TO INFORM OUR FELLOW CITIZENS ABOUT OUR HISTORY BUT ALSO TO ENERGISE AND ENGAGE THEM ON BEHALF OF REFORM AND CHANGE, TO MAKE OUR COUNTRY SAFER AND MORE SECURE.

FOR THIS REASON, WE ARE PLEASED TO HAVE THE OPPORTUNITY TO BRING THE WORK OF THE 9/11 COMMISSION TO THE ATTENTION OF A NEW SET OF READERS. WE COMMEND THE TALENTED GRAPHIC ARTISTS OF THIS EDITION FOR THEIR CLOSE ADHERENCE TO THE FINDINGS, RECOMMENDATIONS, SPIRIT, AND TONE OF THE ORIGINAL COMMISSION REPORT. THEIR ADAPTATION CONVEYS MUCH OF THE INFORMATION CONTAINED IN THE ORIGINAL REPORT. WE BELIEVE THAT YOU WILL FIND THE STORY OF 9/11 A GRIPPING ONE, WHETHER IN NARRATIVE OR PICTORIAL FORM............

THOMAS H. KEAN[4]
CHAIR

THE 9/11 COMMISSION

LEE H. HAMILTON[4]
VICE CHAIR

THE 9/11 COMMISSION

Connect with your Audience's World — *Indirectly*

Pictionary is a family board game. A player in one team is given a card with a word written on it which he has to draw, and the other players in his team have to guess the word on the card within one minute of the 'artist' beginning to sketch. One minute may not seem very long, which is why many people who play this game jump straight in to drawing the object written on their card. So, for example, if the card says 'pepper pot' – the artist draws what they believe to be a perfectly good and accurate representation of the object. For the team, however, it looks like a lighthouse or a candle. Exasperated, the drawer madly begins a new sketch from another angle, from the top, to show the holes in the pot. People shout out, 'rose bowl!', 'pencil holder!' The seconds ticking away, the artist tries again, drawing a salt cellar, and beside it a pepper pot. More wrong guesses as people, now under immense pressure, cast wildly for ideas: 'book-ends!', 'walking sticks!', 'Gog and Magog!' The minute is up and the disappointed artist cannot understand how his teammates could have been so dense.

Run the game again and this time the player sketches a table and the teammates call out 'table!' It is wrong so the player carries on and draws a place setting on the table. All the different objects he draws are correctly identified and called out. He gets to the salt cellar, which he draws as a tiny blob on the centre of the table, and the teammates get it straight away. Then he draws another blob beside it: 'pepper pot!' All the members of his team congratulate themselves for successfully identifying the drawing in under 15 seconds. Beginning with something the teammates recognise and then taking them slowly through familiar objects to the answer was both undemanding and successful.

At a recent workshop, a scientist gave a short presentation to a mixed group of bio-medical researchers. He launched straight into telling us he worked on the cytoskeleton and then the Audience was catapulted into the inner World of the cell and to the molecular details of his work on the cytoskeleton.

He began:

> 'My area of research is the cytoskeleton. I am interested in the septins, a novel component of the cytoskeleton, which are critically linked to ageing, cancer and infections.'

A few of the Audience were with him completely. In the discussion afterwards, several of the group reported that they felt under pressure as they tried unsuccessfully to form a picture in their minds of the World the scientist was describing. One of the members of the Audience suggested he began in a different way – the 'Pictionary way'.

> 'You are all familiar with the cell.'

Immediately we have a picture in our minds of a single cell.

> 'Within that is, of course, the nucleus – the brain of the cell – and then there are the other constituents that do the work like the mitochondria, and these workers are within the cytoplasm of the cell.'

Acknowledge that most in the Audience will recognise the constituents of the cell.

> 'I work on the cytoskeleton, which is also in the cytoplasm and which gives the cell its shape.'

Allow the Audience to absorb the information about the cytoskeleton.

This second version takes people a little more slowly to the World in which the new ideas, the Characters in the drama of septin assembly can be introduced. It acknowledges that some people are familiar with the cytoskeleton and others are not. It brings everyone to the same platform, before starting to introduce new ideas to the narrative. If listeners are preoccupied with trying to work out 'where they are' they will be unable to relax and focus on the information you want to give them. They will feel under continuous pressure to keep up.

Connect with your Audience's World — *Use analogies, metaphors and similes*

A useful way to transport your Audience from their World into your specialist one is to find an analogy – an idea which links the two Worlds.

'You see, wire telegraph is a kind of a very, very long cat. You pull his tail in New York and his head is meowing in Los Angeles.

Do you understand this?

Radio operates exactly the same way: you send signals here, they receive them there. The only difference is that there is no cat.'

Attributed to Albert Einstein; source unknown

Analogies work when an object or a process – familiar to the Audience – has similar attributes or behaviours to an object or process with which the Audience is unfamiliar.

In the 1990s, physicists began in earnest to search for the Higgs boson, an elusive particle which their theories predicted should exist. Physicists wanted millions of pounds to track down this mysterious particle and applied to governments for funds. The Right Honourable William Waldegrave MP was the UK cabinet minister responsible for science at the time. The problem, as he saw it, was that physicists wanted the taxpayer to fund the research, yet the taxpayers had no knowledge of the particle or why it was so important. He therefore set the scientific community a communication challenge. He offered a bottle of champagne (paid for out of his own pocket, apparently) to the best answers to the question 'What is the Higgs boson and why do we want to find it?' The answers needed to fit on one side of A4 paper.

Of the hundreds of entries, five received prizes. The first two paragraphs of three of the winners are included to show the range of approaches to this task. The winning entry by David J. Miller is included in full.

Ripples at the Heart of Physics[5]

by *Simon Hands*

The Higgs boson is an undiscovered elementary particle, thought to be a vital piece of the closely fitting jigsaw of particle physics. Like all particles, it has wave properties akin to those ripples on the surface of a pond which has been disturbed; indeed, only when the ripples travel as a well defined group is it sensible to speak of a particle at all. In quantum language the analogue of the water surface which carries the waves is called a field. Each type of particle has its own corresponding field.

The Higgs field is a particularly simple one – it has the same properties viewed from every direction, and in important respects is indistinguishable from empty space. Thus physicists conceive of the Higgs field being 'switched on', pervading all of space and endowing it with 'grain', like that of a plank of wood. The direction of the grain is undetectable, and only becomes important once the Higgs' interactions with other particles are taken into account. For instance, particles called vector bosons can travel with the grain, in which case they move easily for large distances and may be observed as photons – that is, particles of light that we can see or record using a camera; or against, in which case their effective range is much shorter, and we call them W- or Z-particles. These play a central role in the physics of nuclear reactions, such as those occurring in the core of the sun....

Of Particles, Pencils and Unification[5]

by *Tom Kibble*

Theoretical physicists always aim for unification. Newton recognised that the fall of an apple, the tides and the orbits of the planets as [*sic*] aspects of a single phenomenon: gravity. Maxwell unified electricity, magnetism and light. Each synthesis extends our understanding and leads eventually to new applications.

In the 1960s the time was ripe for a further step. We had a marvellously accurate theory of electromagnetic forces, quantum electrodynamics, or QED, a quantum version of Maxwell's theory. In it, electromagnetic forces are seen as due to the exchange between electrically charged particles of photons, packets (or quanta) of electromagnetic waves. (The distinction between particle and wave has disappeared in quantum theory.) The 'weak' forces, involved in radioactivity and in the Sun's power generation, are in many ways very similar, save for being much weaker and restricted in range. A beautiful unified theory of weak and electromagnetic forces was proposed in 1967 by Steven Weinberg and Abdus Salam (independently). The weak forces are due to the exchange of W- and Z-particles. Their short range, and apparent weakness at ordinary ranges, is because, unlike the photon, the W and Z are, by our standards, very massive particles, 100-times heavier than a hydrogen atom....

The Need to Understand Mass[5]

by *Roger Cashmore*

What determines the size of objects that we see around us or indeed even the size of ourselves? The answer is the size of the molecules and in turn the atoms that compose these molecules. But what determines the size of the atoms themselves? Quantum theory and atomic physics provide an answer. The size of the atom is determined by the paths of the electrons orbiting the nucleus. The size of those orbits, however, is determined by the mass of the electron. Were the electron's mass smaller, the orbits (and hence all atoms) would be smaller, and consequently everything we see would be smaller. So understanding the mass of the electron is essential to understanding the size and dimensions of everything around us.

It might be hard to understand the origin of one quantity, that quantity being the mass of the electron. Fortunately nature has given us more than one elementary particle and they come with a wide variety of masses. The lightest particle is the electron and the heaviest particle is believed to be the particle called the top quark, which weighs at least 200,000 times as much as an electron. With this variety of particles and masses we should have a clue to the individual masses of the particles...

These three winners addressed and answered, in different ways, the question posed by the science minister. All of them made the same assumptions about the Audience: that the reader was already interested in the topic and would give their full concentration to the A4 page. In this case, their assumptions were correct. However, David Miller made further assumptions. He assumed that the person reading the piece would be a time-poor cabinet minister with no degrees in science; that he therefore probably had limited tolerance for dense or detailed explanations. As a result his piece provides some further insights about engaging a more diverse group of readers.

Miller's title, *A Quasi-political Explanation of the Higgs Boson*, succeeds in linking his subject matter directly with the intended reader – the cabinet minister. If that was not enough to have engaged his attention – and it probably was not – the analogy of a former prime minister walking into a cocktail party of political party workers, would certainly have resonated with the minister's direct experience. Immediately, and without any obvious effort on his part, he can visualise the familiar behaviours of people around power and celebrity. Once his imagination is engaged, it is much easier for the narrator to begin to interest him with the behaviour of something new and possibly unknown which acts, in some ways, in a vaguely comparable manner.

A Quasi-Political Explanation of the Higgs Boson; for Mr Waldegrave, UK Science Minister, 1993[6]

by *David J. Miller*

A. The Higgs Mechanism

Imagine a cocktail party of political party workers who are uniformly distributed across the floor, all talking to their nearest neighbours. The ex-Prime Minister enters and crosses the room. All of the workers in her neighbourhood are strongly attracted to her and cluster round her. As she moves she attracts the people she comes close to, while the ones she has left return to their even spacing. Because of the knot of people always clustered around her she acquires a greater mass than normal; that is she has more momentum for the same speed of movement across the room. Once moving she is hard to stop, and once stopped she is harder to get moving again because the clustering process has to be restarted.

In three dimensions, and with the complications of relativity, this is the Higgs mechanism. In order to give particles mass, a background field is invented which becomes locally distorted whenever a particle moves through it. The distortion – the clustering of the field around the particle – generates the particle's mass. The idea comes directly from the physics of solids. Instead of a field spread throughout all space a solid contains a lattice of positively charged crystal atoms. When an electron moves through the lattice the atoms are attracted to it, causing the electron's effective mass to be as much as 40-times bigger than the mass of a free electron.

The postulated Higgs field in the vacuum is a sort of hypothetical lattice which fills our Universe. We need it because otherwise we cannot

(*Continued*)

explain why the Z- and W-particles which carry the weak interactions are so heavy while the photon which carries electromagnetic forces is massless.

B. The Higgs Boson

Now consider a rumour passing through our room full of uniformly spread political workers. Those near the door hear of it first and cluster together to get the details, then they turn and move closer to their next neighbours who want to know about it too. A wave of clustering passes through the room. It may spread out to all the corners, or it may form a compact bunch which carries the news along a line of workers from the door to some dignitary at the other side of the room. Since the information is carried by clusters of people, and since it was clustering which gave extra mass to the ex-Prime Minister, then the rumour-carrying clusters also have mass. The Higgs boson is predicted to be just such a clustering in the Higgs field. We will find it much easier to believe that the field exists, and that the mechanism for giving other particles mass is true, if we actually see the Higgs particle itself. Again, there are analogies in the Physics of Solids. A crystal lattice can carry waves of clustering without needing an electron to move and attract the atoms. These waves can behave as if they are particles. They are called phonons, and they too are bosons. There could be a Higgs mechanism, and a Higgs field throughout our Universe, without there being a Higgs boson. The next generation of colliders will sort this out.

David J. Miller

Risk — *Where Worlds part*

In the *Audience* chapter we briefly touched on the World of risk – a place where the laws of probability and statistics apply. Even for people trained in mathematics, risk is a non-intuitive concept and so is difficult to visualise. There are, however, exceptions to this.

In this example from *Reckoning with Risk*, Gerd Gigerenzer[7] gives two different ways of explaining the results of mammography to women who test positive. The first way uses probability. See if you can work out the answer to the question, 'What are the chances that a woman who tests positive actually has breast cancer?'

> The probability that a woman of age 40 has breast cancer is about one percent. If she has breast cancer, the probability that she tests positive on a screening mammogram is 90 percent. If she does not have breast cancer, the probability that she nevertheless tests positive is nine percent.
>
> What are the chances that a woman who tests positive actually has breast cancer?

The second method of explaining the results uses natural frequency. It is not the same as probability, but it also has validity.

In the first short sentence the reader is asked to enter a World by thinking of 100 women. It is a simple and easy instruction and everyone can do it. With that image in their head, readers are given information which adds to the image without fighting it. This explanation is straightforward, direct and does not require a level of interpretation which the introduction of words like probability and percentages make necessary.

> Think of 100 women. One has breast cancer, and she will probably test positive. Of the 99 who do not have breast cancer, nine will also test positive. Thus a total of ten women will test positive.
>
> How many of those who test positive actually have breast cancer?

Some researchers look at these two examples and bring up the shortcomings of the natural frequency example: it does not, for example, mention false negative results in the mammography. That information is important, and if it is relevant to the particular Audience, it can be added separately. People need time to absorb new information and they find it is easier to do so if the fresh ideas come one at a time in an orderly fashion.

It is not only lay members of the public who are influenced by how the World is introduced with the information presented. In an experiment, those who decide whether criminals should be released from prison were presented with expert risk assessments of the likelihood of the prisoners reoffending.[8,9] They were asked to read all the information and decide whether or not to approve the release of the prisoners.

The facts were identical; their presentations differed in only one respect. When the risk was presented as a probability, almost one in two of the professional decision-makers approved the prisoners' release. When the risk was presented as a frequency only one in four voted for their release.

The same effect has been found with doctors assessing the severity of illnesses. Frequency is more vivid; it invites Audiences to think of individuals, to see real people and not abstract percentages.

Enabling your Audience to *visualise* your World is an important ingredient in communication – essential for unfolding narratives.

Visionary Worlds

Creative professionals, from advertisers to novelists, all aspire to create imaginary or real Worlds where their Audiences will be receptive to their ideas and stories. For advertisers it is their sole purpose of existence. How else are they to make profitable connections with an Audience? Good advertising copywriters for hotels and holidays, for example, are able, through pictures and words, to give their readers a sense of what it will be like to be a guest. They succeed in transporting their readers – willing participants – into the World they have described. At least that is what the advertisers hope happens. If the readers are not looking for a holiday, they are unlikely to give the advertisement a second glance.

There are, however, individuals who are able to create visionary Worlds which have extraordinary power to connect with their Audience. Successful leaders, from all occupations (including science and engineering), provide their followers with a vision of a World which is more than simply a description of what it is or will be like. They provide a credible and inspiring sense of what it will feel like being enmeshed in the World. Visionary leaders construct a partial World in which their followers feel important and valued; somewhere they can belong and take responsibility. It is no mean feat. Martin Luther King in his *I Have a Dream* speech and Winston Churchill in his wartime broadcasts both demonstrated this skill. Choice of words, imagery and, for the spoken word, a passionate delivery, connect with individuals and provide a partial sketch of the World, which gives the Audience space to imagine the rest, including their own place and actions within it. Visionary Worlds can shape thinking and inspire action.

This is the visionary World which Walt Disney created for his staff. What is the visionary World for your workplace?

> The idea of Disneyland is a simple one. It will be a place for people to find happiness and knowledge. It will be a place for parents and children to share pleasant times in one another's company, a place for teachers and pupils to discover greater ways of understanding and education. Here the older generation can recapture the nostalgia of days gone by, and the younger generation can savor the challenge of the future. Here will be the wonders of nature and man for all to see and understand. Disneyland will be based upon and dedicated to the ideals, the dreams and hard facts that have created America. And it will be uniquely equipped to dramatize these dreams and facts and send them forth as a source of courage and inspiration to all the world. Disneyland will be something of a fair, an exhibition, a playground, a community center, a museum of living facts, and a showplace of beauty and magic. It will be filled with accomplishments, the joys and hopes of the world we live in. And it will remind us and show us how to make those wonders part of our lives.[10]

In the next chapter we look at your relationship with the World you have created for the Audience: you, as a Character within the narrative, driving the story.

Questions and Ideas to Help Stimulate Ideas for Introducing your World

Here are some questions to stimulate ideas for new ways to introduce your subject to an Audience. If the Audience is very unfamiliar with your work, you may want to experiment with some radical new ideas.

1. Imagine you are a tour guide, taking your Audience on a trip into your World. Describe it as you would to a group of tourists who are on a coach trip or a walking trip through your World. (If this seems ridiculous, bear with it and when you have finished, think about what you can take from your piece which you can adapt and use.)
    - Where is this World? What are the most prominent landmarks or features?
    - What does the environment feel like?
    - Why is this World interesting?
    - How do you get to it, how is it connected to other Worlds?
2. Take three pieces of A4 plain paper and draw your World as a 'cathedral', a 'wall' and a 'brick'.
    - Describe the images in words.
3. Use a technique mentioned in the questions section of the *Lure* chapter: take one of your drawings and force connections between its features and features in the everyday World of your Audience. Use this to try to generate analogies to describe your World. Another way to do this exercise is to take a set of random pictures from magazines or postcards and force a connection between them and each of your three drawings. Most of your attempts will probably lead to ideas which go nowhere, but when you have finished, look for the one or two ideas which you can make work for you.
4. Until now you have taken an overview of your World. You are looking down at it. In this exercise the idea is to take the viewpoint of a feature or a landmark within the World.
    - So, how would you describe the World from a completely different perspective? For example, if your World is one of disease or cell biology, how would it look from the point of view of the disease organism – say a mosquito, a virus or a cancer cell bent on immortality?

5. Can you introduce a historical perspective to your World?
    - Are there stories of pioneers or personal heroes from your specialist area which could provide ideas for setting a context? Can you tell one of these to introduce your World?
6. Is there a recent event or story which you can use as a launch pad to interest people in your World?
7. If you were making a film trailer for your narrative, what would you include? Adapt this to a written or verbal introduction to your World.

5

Character: You! – the Audience's guide, in search of a prize

> 'Scientists have changed our lives more dramatically than television stars, statesmen and generals, but the public knows little about them beyond the caricature of the soulless hermit toiling away at abstruse problems that he cannot explain except in incomprehensible gibberish.'
>
> Max Perutz[1]

If you are a science researcher it is a given that you will both attend conferences and speak at them. At first almost everyone finds this an ordeal, more so if English is not a first language. Agnieska, a Polish post-doctoral student, is typical. She had attended several conferences, seen speakers she knew – and many she did not – deliver papers and watched the way that Audiences reacted to them. She had been present when senior scientists in the Audience had 'turned' on speakers, questioning them fairly robustly. She had sympathised with the speakers struggling to reply. Agnieska was resourceful. She had managed to avoid delivering a conference paper during her PhD but, when she became a post-doctoral student, her supervisor insisted that she attend and deliver a paper on her work at an international conference in the USA. She tells an amusing and passionate story of how she went to her supervisor several times, each time with a different and more pressing reason why she could not go, and why perhaps someone else should deliver her paper. He was having none of it. Finally, she resorted to begging him not to make her go. She told him why she did not want to talk in public: her English was simply not good enough. Her supervisor told her that he had always understood that that was her concern, but that she should take this opportunity for two reasons. First, if she wanted to make a career in science, she had no choice but to present her work at conferences in English. Second, the sooner she started, the more practised and therefore better she would become. Finally he told her that because she was young (and looked even younger) the Audience would be sympathetic. If she delayed the inevitable until she was older, Audiences might not be so forgiving.

It takes courage and practice to be comfortable speaking in front of your peers and seniors who are making judgements about you and your work. If English is not your first language it is doubly difficult.

This chapter is about Character: your Character and its relevance to your communications. We shall look at the role of Character in narrative and the Audience's relationship with the Character. We shall also look at ways to make communication less of an ordeal: being in control of your content, and being in control of its delivery.

There are some presentation tips to give you confidence, enhance your authority on a stage and help deal with nerves. Finally, at the end of the chapter, there are a series of questions to help you devise appropriate ways to allow the Audience to understand your connection with your narrative World, and to want to be there with you. In addition there are some references to material to help improve written English and writing style, which is outside the scope of this book.

You are a Character in Your Communication

When I was in my early 30s, I made my first half-hour television documentary film, *Another Little Drink Won't Do Us Any Harm.* (It was in the days before digital TV planners and snappy titles became the norm.) The programme was about the perils of drinking, particularly underage drinking: higher rates of accidents and crime, deaths from fire, deaths on the roads and ultimately potentially premature death from a horrible disease. The backbone of the film was a number of interviews which I had conducted with school-age drinkers and their parents. This documentary was made for the popular BBC science series QED, which was broadcast in the UK between 1982 and 1999. The commentary over the illustrated sections between the interviews was, by tradition, read by a narrator – usually an actor in a serious and slightly detached voice. The idea was, I now think, that the seriousness bestowed an authority on the material, and that the detached voice gave an impression of neutrality. Both of these attributes (objectivity and neutrality) I associated with the scientific way of writing, where the method, results and conclusion are presented as factual information. I believed that the Character of the author was neither relevant nor revealed in scientific papers.

This rather warped idea of the role of the author in a television programme was challenged when my editor suggested that I be the narrator and read the commentary myself. I looked at my script which I had happily written for a 'voice of authority' to read and was aghast. 'There is no way that I am reading this!' I thought. It was wordy, contained too much information and at times verged on the pompous and self important. I suddenly saw it as it really was: muddled and a bit dull. The realisation that I would be the voice of my commentary made me view it in a completely different way. For a start, as the narrator I could not hide behind the objective and neutral words. I had to face up to the responsibility of being the author or Character of the piece, and it jolted me out of my happy but misguided view that authorship is neutral.

So, what was going through my mind when I wrote my first few drafts? I think it is very similar to what goes through many scientists' minds when they are writing a scientific paper or a thesis: research material takes time and effort to gather. Researchers invest heavily in the process and care deeply about the outcome. They believe that their work and their results are important – or should be – and they, as the Character in their own narrative of their career, want their paper to have significance so that others see its worth. The underlying purpose of the communication can become distorted. Worthiness replaces simplicity and clarity. The words on the page are imbued with gravitas and detachment to impress readers with the author's knowledge, rather than to explain simply and clearly what was done, why and how, and ultimately what happened and what it means.

Authors are not outside the World of the story, and neither are they disinterested observers in the World. They are a Character, albeit unseen, in the World in which the narrative will unfold. It is the Character who has asked the research question, done the work, collected the data and written it up. At least as far as readers and listeners are concerned. The great dilemma for scientists as communicators, however, is that the personality of the author is totally irrelevant to the science.

Be in Control of Your Content

Science is a form of knowledge based on empirical evidence. It is the weight of evidence which carries the day in science, not the force of an individual Character's personality. The biggest favour that you can do for your science, however, is to present your content in a way that others can understand. Some of the scientists who come to my workshops tell me that their professors positively discourage clear writing. I found this hard to fathom until I had a conversation with a brilliant communicator, the Manchester University graphene scientist (and Nobel Prize winner), André Geim. He expressed a view that the better a paper is written, the more difficult it is to get published. The reason for this being that if reviewers have no difficulty understanding the paper and its arguments, they form clearer opinions of its worth, and may disagree with it. However, he told me that the reverse is true of grant applications. Better written grant proposals are more likely to be funded than badly written ones. When money is involved, reviewers want to understand how it will be spent. Geim did point out, however, that published papers which are *well* written tend to be widely cited, even more widely read and their authors admired by grateful students.

My view is rather idealistic. I believe that people should always communicate with the intention of having their meaning understood, even though this may not be easy.

In many areas of science, the terminology is unfamiliar and the precision with which things need to be described is exact; these factors can make reading papers, posters and grant applications challenging. Yet just because the terminology is challenging, it is no reason to make the whole communication equally so. Sentences which are written simply and express a single thought are easy to read. Sentences which build on previous sentences to add new information are also easy to read. An article constructed with easy-to-read sentences gives an impression of an author who is a Character in control of his or her World; a protagonist who gives the Audience confidence.

Conversely, this sentence which, in order to illustrate a point, is going to attempt to describe a single idea with clarification so that it includes, to ensure lexical difficulty, sub-clauses, which, in addition to adding to the length of the sentence serves the gratuitous purpose of endowing it with burdensome adjectives and adjectival clauses, notwithstanding complex definitions which satisfy the author's desire for displaying genius and rare expertise, but creates in the reader, who in all likelihood will not make it to the end of the sentence before needing to stop to consider what it is that the author is attempting to define, a headache at the very least.

> 'I would never use a long word where a short one would answer the purpose. I know there are professors in this country who "ligate" arteries. Other surgeons tie them, and it stops the bleeding just as well.'
>
> Oliver Wendell Holmes, Sr[2]

There are many useful books and websites on how to write scientific papers and theses.[3] In addition, guides to writing plain English give advice on writing style.[4] In essence, the purpose of your writing is to communicate with a reader accurately and clearly. Every writer will tell you that writing is not easy. It takes many attempts, many drafts before the piece is as clear and as accurate as it can be. When you succeed, your readers feel satisfied that they have been guided by a Character with skill and knowledge.

A word about jargon

When jargon is used as a shorthand amongst a small group of colleagues it is perfectly acceptable because everyone understands its meaning and it speeds up communication. Once the small group becomes a larger one and includes individuals or groups of people who are not familiar with the shorthand it becomes rude to use it. It is the equivalent of a group of international researchers who have different mother tongues but share a knowledge of English, and use this commonly understood language when they are together. They will use their first language with their colleagues from home. The moment someone who does not speak their language comes into the room, however, they revert to English. It indicates a Character trait. It is certainly more difficult to speak English, it requires more effort; nevertheless they do so to be inclusive and polite.

Jargon is pervasive across all specialists' communication; it is not an issue for scientists alone. Golden Bull Awards are given each year by the Plain English Campaign for writing which is not plain English.

This was one of the 'winners' for the 2012 award:

> 'A unique factor of the NHS Cheshire Warrington and Wirral Commissioning support organisation is its systematised methodology for project and programme management of small, medium, large service re-design and implementation... Building in equality and risk impact assessments the options are taken through a process to arrive at the content for an output based specification and benefits foreseen as a result of the implementation.
>
> The service is inclusive of full engagement with Clinical Commissioning Groups who direct at decision-making points how they wish the proposal to be deployed (re-commission, de-commission or changes to current services/providers), and lastly an implementation team who see the service redesign through to evaluation and benefits realisation.'[5]

Reading written jargon requires effort and several read-throughs to interpret the intended meaning, even when English is your first language. Jargon in a talk, where there is no chance of going back, often means that the Audience is left struggling or lost.

In 2012, Alan Duncan, the government minister at the Department for International Development rebuked his staff for using 'language that the rest of the world does not understand.'[6] Alan Duncan was referring to the jargon which riddled many of his department's official documents. He wanted to stamp it out. In a memo which was published on the internal website (and in some newspapers) he asked his officials to think how a member of the public would interpret any information. 'All our communication must be immediately explicable to the non DFID reader. Clear language conveys clear thought. Its poor use suggests sloppy thinking.' Could that accusation also be levelled against scientists, researchers and other specialists?

Some years ago Martin Gregory wrote a piece in the science journal *Nature* entitled 'The infectiousness of pompous prose'.[7] He included the following example:

> 'As the relevance of intestinal immunity in diarrhoeal disease relates to the possibility of developing effective immunisation programmes for the control of gut infections, this review will focus on insights into the functioning of the immune system particularly relevant to this goal.'[8]

In other words, as Martin Gregory paraphrased, 'This review will focus on aspects relevant to vaccine development'.

At the time I read this, I picked up the current edition of *Nature*, which was in the office, and found the following:

> 'The springtime reduction of ozone over Antarctica is now well established observationally and the processes involved are quantitatively understood theoretically. Whether similar processes could operate over the Arctic to produce massive reduction in the ozone abundance remains uncertain.'[9]

In other words: 'We know how and why the ozone disappears above Antarctica in spring, but we do not know whether the same thing could happen in the Arctic'. You do not need to be a scientist to translate this convoluted sentence into one which anyone can understand. Its construction gives an impression of an author who is desperate to impress; or perhaps a Character who is overwhelmed by their content and not in control of it.

> 'Make things as simple as possible but not simpler.'
>
> Attributed to Albert Einstein

When you are presenting to members of the general public many of the words which scientists use to describe particular features in a scientific context have other meanings in everyday language. This table gives a few examples; you may well have many more examples from your own specialist areas.

Word	Scientific context	Everyday context
Host	Home to a parasite.	Person giving a party.
Genesis	Creation.	1st book of the Old Testament or a rock band from the 1970s.
Field	Sphere of activity.	Piece of land.
Energy	Power from physical or chemical resources.	Strength and vitality for physical or mental activities.
Base	An alkali.	A foundation or starting point.
Barn	Unit of area (10^{-28} m^2) in particle physics.	A farm building.
Cell	A device which generates electric current, or the smallest structural and functioning unit of a single organism.	A small room where prisoners are locked in; the bedroom of monks and nuns; or the American term for a mobile phone.
Transcription	The process by which genetic information is copied.	A written or printed representation of something.

Be a Hero

The Author as a presence may not be apparent in a research paper or thesis, but the scientist who speaks at a seminar or in a public arena is, at least as far as the Audience is concerned, a Character in their own narrative.

Scientists who communicate well are able to bring their Audience fully into their World; they can take on the role of *hero*. A hero in classical storytelling is someone who embarks on a quest to reach a goal which involves struggle and self-sacrifice. In the process the hero learns and grows; something that the Audience identifies with.

In science, the researcher hero embarks on a quest for knowledge in an important area of science. Over a career they will experience failures, successes and a fair amount of struggle. They will grow as a result. How they adapt to the different situations they find themselves in reveals something of their Character. Every time they make a decision to take one path or another they reveal something about what is important to them and therefore who they are.

> 'It is our choices, Harry, that show what we truly are, far more than our abilities.'
>
> Professor Albus Dumbledore to Harry Potter[10]

For example, if you choose to stay with an experiment rather than go away for the weekend, it implies self-sacrifice, a key hero attribute. If however you had been ordered by your boss to stay with *his* experiment, and you were too fearful to ignore him, it might suggest something different about you. Whatever the actual situation, the Audience likes and needs Characters in narratives. They need them because they want to experience, albeit vicariously, the World in which the narrative unfolds. They experience the World by identifying with the protagonist: the hero. They want to learn about themselves from the actions of the hero. They ask themselves what they would do in the same situation.

Form a Relationship with Your Audience

Good speakers form a relationship with the Audience by revealing something about themselves with which the Audience can identify.

The revelation can be a story or anecdote which is relevant to the main communication. Thus, one research scientist on a workshop I ran introduced her research into Alzheimer's disease with a story of her visits, as a child, to see her grandmother in a nursing home. The story served the purpose of introducing the Audience to the World of the talk, it also gave the Audience an insight into the motivations of the Character of the researcher, revealing something of her values with which they could sympathise.

Personal stories do not need to be personally sensitive or private. They simply need to give the Audience a desire to engage with you and your World: you and your quest. They act as a bond or a glue to connect you to your Audience, and sometimes to your topic as well.

I was once asked to give an after dinner speech at a public service leadership conference. I wrote the first draft without making any overt connections between me and the Audience. My first sentence read:

> 'I thought I would tell you some stories from my time at the BBC which reflect how attitudes to audiences have changed in the last 30 years.'

For my next few drafts I forced links between the BBC and the public sector by asking myself what they had in common: I went to the conference with this as my first paragraph:

> 'The BBC is both different and similar to other parts of the public sector. It spends money rather than makes it. Its consumers have high expectations, its critics are articulate and vocal. Leaders are rarely out of the spotlight. Small mistakes are leapt on with the aim of drawing blood, and big ones become bandwagons for politicians to score points.'

While I was at the conference, I listened to a few of the speakers. Two of them in separate talks mentioned, in reply to questions, some issues with a particular newspaper – the *Daily Mail*. I leapt on this and incorporated it into my opening paragraph. I ended my introduction that evening with the sentence, 'And finally: the *Daily Mail* is not our friend'. These few words made the difference between a moderate connection with the Audience, that is, people nodding quietly in agreement, to a strong connection, that is people laughing out loud. The listeners – my Audience – had accepted the connection between my World and theirs. The extra sentence demonstrated our shared experience with the *Daily Mail*. I now had their full attention.

> 'I have an old photograph in my office. I am 30 years younger, on skis, traveling at about 60mph and airborne. It is on my wall to remind me of the dangers of complacency. Back then I thought I was a pretty good skier, but this was my first (and only) attempt at a natural ski jump, and I was in new territory with no prior training or instruction in how to manage it. The picture shows me struggling to maintain balance and equilibrium, and recognising that the techniques which had carried me through under the old familiar circumstances of the *piste* was not well adapted to this new and unfamiliar terrain.'
>
> Professor Peter Slee[11]

This paragraph was written by Peter Slee when he was deputy vice-chancellor of the University of Huddersfield. It was his first paragraph of an article about strategy in higher education. The old photograph was not shown but the author, in words alone, managed to tell a short story about himself and use it as an analogy with which to introduce the World of his topic. He connected himself to his World, and he connected himself to his Audience through a story of his youth which gave them an insight into an aspect of his Character.

Having good content – that is having a compelling narrative – is only part of what is necessary for good communication. The other part of the equation is its delivery. Let us now look at delivery when the content is spoken. For this we need to think of giving a talk or a presentation as a performance.

Be in Control of Your Performance

Researchers who shuffle onto the platform and stand huddled beside the overhead projector clutching notes, and looking backwards at a screen while mumbling in a way which suggests they would rather be somewhere else, do *not* inspire confidence. They look like victims who are not in control of their destiny; they generate a mixture of dread, pity and irritation as the Audience braces itself for an uncomfortable session.

Now think of senior leaders with natural authority and charisma – someone like Nelson Mandela or Aung San Suu Kyi. They have a way of walking, talking and interacting with an Audience which positively exudes confidence. They have what actors refer to as 'an aura', a stage presence. So, too, do good speakers; they have a way of looking, standing and speaking which communicates to everyone who is watching and listening that they are in control and in charge. So what do they do?

I once spent a couple of weekends in London learning improvisation techniques from Deborah Frances-White, comedian and co-author of *The Improv Handbook*[12], who had been taught by the improvisation maestro, Keith Johnstone.[13] During these weekends[14] we played a game where one person was asked to act as the seller of a flat, the other the buyer. The idea of the game was that the two on stage were both asked to try to be the higher status individual – the person with greater social stature. As they acted out the scene, Deborah would stop the action and ask the group who, at that point, was the higher status individual. Surprisingly it was always clear. We knew from watching, who was in control of the social situation. The lower status individual left the stage, another came on and the game began again.

I learned a lot about how to perform on stage as a speaker from these weekends. Simply watching people in small groups – bosses and workers, professors and PhD students – where their social standing is uneven can give many clues about how to become master or mistress of the space – a skill which can be useful when you are performing in front of an Audience.

One of the characteristics of individuals with 'low status' is that they defer to those of higher statuses by casting quick glances in their direction while avoiding eye contact. This can be most obvious at seminars where the most senior person, the professor, looks at the speaker while the Audience members and the speaker, all of lower status, constantly glance towards the professor. They are looking at the 'leader', they want reassurance. When you are giving a presentation, the Audience want you to be the leader and the person in control. They want you to have high status. It is reassuring for them.

There are a number of ways to spot the high status individuals, the alpha males and females among the human and ape populations.

- High status individuals fill the space with their presence. They stand comfortably tall on two feet with their underbelly exposed.

  Try standing upright but with your lower belly slightly protected by hunching over it. Then stand erect and expose your underbelly. What do you notice? Does it feel different? It also looks different. If you stand in front of a mirror and try it, you might notice you look less vulnerable, but feel more so, when you expose your underbelly.

  Lower status individuals hunch very slightly to protect their underbelly, a weak spot in anatomical terms. If we pass someone who has a higher status than us (either real or imagined), unconsciously we want to give them more room. I tried this with some people in a workshop, asking them to imagine I was the director of their institute; one researcher unconsciously bowed slightly as he passed. People laughed, but they recognised the situation.
- High status individuals hold their heads erect exposing their jugular vein in the neck – another anatomical weak spot. They don't twitch or make unnecessary movements. They don't use their hands to cover their mouths or touch their face.
- High status individuals who look in control, move purposefully, smoothly and deliberately. They make full eye contact with people they address and they keep their head still when speaking.

Be Master or Mistress of the Space

Individuals with a high status aura	Individuals with a low status aura
Occupy the space – fill it with their presence.	Stand slightly hunched.
Move purposefully, (smoothly, deliberately, unhurriedly).	Move hurriedly, twitch, make unnecessary movements.
Make eye contact with members of the Audience, give attention to all parts of the Audience.	Avoid eye contact.
Keep head still when speaking.	Move head while speaking.
Speak in complete sentences, sometimes use long 'ums' and 'ers' as way of filling the gaps, preventing interruptions.	Use short, 'er' at start of sentence.
Keep their hands well away from their faces.	Use their hands to touch their face.
Can ignore others.	Look for approval with quick glances at higher status individuals.

You can try watching people to spot these postures and to decide how many you want to adopt to enhance your presence on a stage.

The most important act for you as a Character giving a talk is to make a connection with the Audience. Eye contact is the key, even if the arena is vast and the Audience a long way away.

One of the exercises I do in my workshops is to ask researchers to walk on the stage, put down their papers and make eye contact with the Audience for a full three seconds before speaking. Almost everyone finds it really hard to do. The three seconds can feel like an eon; the temptation to cut it short overwhelming. People have the urge to start talking as they put their papers down, or to look at the technology and hurry themselves. However, it is the one exercise which I ask people to do until they manage to last out the full three seconds. Some attempt it a dozen times, usually with much gentle amusement from the Audience, who all take part. The reason that

I put mild pressure on people to take part is because of what happens when they succeed – as they eventually do. Not only do they look more confident and in control, they clearly feel stronger too. It is not unusual at the end of the session for a string of people to ask to have another go.

If you can, keep eye contact with an individual for a second, then transfer your eye contact to another person in another part of the Audience for a second, and finally to a third person for a final second before speaking. Your entire Audience (and not only the three people you have made eye contact with) will feel a connection with you. If the auditorium is so large that it is hard to single out individuals in the Audience, look at a block of the Audience as if they are an individual and they will believe that you looked right at them. Make eye contact and you begin to form a relationship. Mastering this simple technique will connect you to the Audience; it will give you – and them – confidence that you are in control. Try it.

If someone starts a conversation in the Audience, a low status individual will try to ignore it, or perhaps talk a bit louder. As good teachers and high status individuals know, it is stillness which gets attention. If you stop talking, remain still and make eye contact with the disrupter, then the whispered conversation will almost certainly cease. Of course you may not want to make such a pointed move. An alternative is to pause (always welcome in a talk as it allows people time to digest what has been said), reach for a glass of water, and take a drink, to make the rebuke less obvious.

And Finally

Performance makes physical demands on speakers. Just as footballers and opera singers prepare before they enter their performance arena, speakers too should prepare. Even small preparations can help transform your nervousness to 'anticipatory excitement.' At least, that is how I like to look at it.

Some tips which I have found helpful are listed below.

To feel comfortable in body

- Breathe in slowly, drawing the breath through the nose down to the tummy and as far as possible below the belly button. (You can practice this by lying on the floor with a book or a bag of sugar on your stomach. When you breathe in, the book should rise; it falls as you breathe out.) Breathe out slowly allowing the jaw, shoulders and arms to go loose and limp. Repeat three times.
- Tense jaw, shoulders and arms; hold for 2–3 seconds and then let go.
- Stand and breathe deeply (taking air right down to below the belly button again) for a few breaths and feel relaxed. Then reach the arms out to the side, stretching – as if trying to touch walls on either side. Relax shoulders and stretch. Reach arms out in front and stretch further, feeling the muscles in the back and arms. Then reach arms upward to the sky, reach, reach. Repeat the cycle, three times.
- Roll the shoulders back, once, twice, three times, slowly. Then forward three times slowly. Shrug the shoulders up and release three times.
- Take in another long slow refreshing breath. And another until the tension is released.

A brisk walk can have the same effect as a series of physical exercises, and often they are the opportunity to do the voice exercises too.

To prepare the voice

- I try to touch my chin and my nose with my tongue.
- Then move the tongue from side to side towards each ear.
- I massage my jaw.
- I take a deep slow breath and with jaw relaxed let the air come through my lips so they tingle as I hum 'mmmmm'.
- I have a few tongue twisters which I say very slowly and clearly, enunciating every syllable in an exaggerated way.
- I then repeat them as fast as I can.

Round the rugged rock the ragged rascal ran.

I slit the sheet, the sheet I slit, and on the slitted sheet I sit.

A skunk sat on a stump. The stump thought the skunk stunk. The skunk thought the stump stunk. What stunk the skunk or the stump?

Billy Button bought a buttered biscuit. Did Billy Button buy a buttered biscuit? If Billy Button bought a buttered biscuit, where's the buttered biscuit Billy Button bought?

I find these voice exercises help my fluency. I speak more clearly and am less likely to stumble over words.

Many people do not bother with limbering up before giving a talk. If I am giving a presentation in a lecture theatre or to a large group of people I always prepare physically as well as mentally.

If you feel alert, if you have control of your voice, and if you are not holding tension in your upper body, then you will be more likely to feel in control of your performance and more likely to enjoy it. It gives you the confidence to deliver your narrative with *crackle and fizz.*

In the next chapter we turn to what, arguably, is the most important part of a narrative: The Big Hook.

Questions and Ideas to Help Stimulate Ideas for Introducing Yourself as the Character

1. Link yourself to your World
   - Is there a picture or slide which will link you to the subject of your talk for your Audience? Is there some physical object which you can show and describe to your Audience to provide a link between the subject of your talk and you the Character?
   - How did you decide on your field of study, or a particular course of action? Is it relevant? Is it an interesting story to share with the Audience?
   - Think of big decisions or moments in your life. What do they reveal about what is important to you? Is there one which could be relevant to your talk?

- Is there someone who inspired you to do research: a famous scientist from the past, one from the present, a teacher, a relative, someone you know? Could an anecdote about this person provide you with a link to the World of your presentation?

2. *Be in control in your World and your Characters*

- Is your World one which has living creatures? Could any of these creatures take on the role of the main Character? Try telling the story from their point of view. Let them introduce their World with you as a secondary Character.
- Look at your draft script and filter for jargon. Go through each sentence and underline words which an intelligent 14-year-old is unlikely to be familiar with. Is there an alternative more common word which would work as well? Once you have changed all the words which have simpler alternatives, look again at the words which do not. Ask yourself which of these still underlined words your Audience may be less familiar with than you. Define those words for your Audience the first time you use them.

6

Big Hook: a puzzle which drives the narrative

Today might look like any other day in the lab. Only it isn't. It is Thursday, and the science journal publishing your paper will arrive this morning and you are quietly joyous. Walking up the stairs, more breathless than usual, you hurry down the corridor leading to the lab. Your senses are heightened in anticipation. Through the window of your lab you see one of your colleagues is leaning against the bench reading a magazine. As you slide open the door, he looks up, his expression freezes for an instant. He extends his hand, holding the journal, and says in an unusually serious voice, 'I think you better read this'. The page is open at your paper. There is its heading, you scan it quickly. And then you notice. Your name is not there, not anywhere on the paper.

In narrative terms, this last sentence is the Big Hook puzzle. If I have done a half decent job in setting it up you are likely to be more than curious about what this means. Something is clearly amiss. The Big Hook has upset you and you need answers, so in all likelihood, you will be rapidly searching for solutions, assessing possible explanations. Was it an error at the publishers? Was it removed by the lead author? It must be a mistake, mustn't it?

Now compare with this account which has no obvious Big Hook:

Today you are excited because your paper on detection of genetically modified micro-organisms is coming out in the *Micro-biologica* journal and you are looking forward to seeing it. When you get to the lab, your colleague who works in the same field, but on a different organism – one which has been genetically modified in nature – is reading it. As soon as you walk through the door, however, he hands the journal to you, informing you that you should read it. You scan the article which looks good in print. You put it aside to look at it further at lunchtime.

This account has a sense of a story. The Audience is accumulating various pieces of coherent information, assuming from their experience of story that they are going to become relevant later on. It is presented in a slightly detached way. Far better to make the Audience feel they are present, within the scene. (For example, using what people actually say, rather than what they reportedly said is more compelling for an Audience. As Mark Twain famously said, 'Don't say the old lady screamed, bring her on and let her *scream.*'[1]) There are a list of actions which follow one from another but there is no obvious Big Hook. No narrative ingredient which draws in an Audience to engage with a puzzle.

The Big Hook is so called because that is exactly what it is: it hooks the Audience – big time – into a narrative. It gives them a reason for wanting to follow a story.

In fiction, the Big Hook delivers a puzzle which the Character within the narrative, and thus the Audience, cannot ignore. For the Character, the puzzle is unsettling, a powerful motivator initiating action to find ways to remove a discomforting situation.

In romantic fiction it can be the hero making eye contact with someone across a crowded room. In a thriller the Big Hook for the Audience might be a suitcase swapped with another at an airport. In a television series such as *The Apprentice,* where individuals compete over several weeks to be sponsored by Donald Trump or Alan Sugar, the Big Hook in each episode is who will be fired in the boardroom this week. In real life, it can be a person aware, for the first time, of a lump under their arm, an experimental result that does not fit with expectation, a researcher with a hunch, a president challenging his nation's scientists and engineers to 'commit itself to achieving the goal, before this decade is out, of landing a man on the moon and returning him safely to earth.'[2]

All of these Big Hooks will, in different ways and for different reasons, cause Characters within the narrative to take action to meet their challenge, deal with their problem – or opportunity – and fulfil the goal they have set themselves. This, for the Audience, who are proxy witnesses to these events, is where the story begins.

Once the Big Hook is introduced into the narrative, the Audience understand that actions will follow as the Character – intent on finding answers to the puzzle – makes his way through the Plot.

In this chapter we shall look at examples of Big Hook puzzles to show that it is 'gaps in knowledge' which motivate individuals and Audiences to search for answers. In science, Big Hooks set a direction for research; they define each research challenge. We shall show that Audiences only engage with Big Hook puzzles when they have sufficient reasons to care about their relevance and importance.

Finally there are examples and questions to help you find an appropriate Big Hook to be the engine for your different communications.

It is All in the Gaps

My husband David is a scientist and a football fan. He is an expert in both. Experts have certain characteristics which I shall describe in terms of the football. He has played the game and followed the fate and fortunes of his home team, Manchester United, for most of his life. He has a deep understanding of both football and the club. Over the years he has devoured every scrap of information relating to his team, players and managers, so that now no-one could doubt his expertise. He has a breadth and depth of knowledge which only other committed football fans share.

On a match day, my husband will read the pre-match reports in various newspapers. If the game is shown on television, it will be turned on half an hour before the game begins to find out who is playing and who is 'left out'. The ninety minutes of actual play will be followed closely, often with the sofa dragged nearer to the television. The pundits' comments at the final whistle are listened to intently with quiet agreement or noisy disagreement. Next day all the post match reports will be read and digested fully. This is the kind of behaviour which is typical of experts. Experts are experts because they know more than most others about their subject. They acquire their knowledge through many different avenues – from their own experience, from seeking out information in books, magazines, web articles, blogs, attending conferences and in discussion with other experts.

Experts, whether football fans, engineers, accountants, lawyers or scientists, share a strong motivation – some might call it an obsession – for filling the gaps in their understanding of their chosen expertise. The more they know, the more they want to know. The apparent obscurity of the information, or the narrowness of it, is not usually an issue. Their tolerance for hearing the same things repeatedly is developed. (This is useful if one is giving talks to the same people on a regular conference circuit.) People will happily sit through a presentation waiting for any new pieces of information – the something they did not know – which fills a small gap in their understanding. Experts will scan a paper or abstract actively looking for new ideas or information. They will then read the paper more slowly to digest every last morsel of intelligence. If there is an editorial or other commentary on the paper, that too is examined and mulled over.

> If you are talking to a group of colleagues, fellow experts who have heard you talk before, and if you are speaking, and not reading, you are more likely to add to your Audience's understanding. Why? When you speak without a script, you are very unlikely to say exactly the same thing using the same words as before. Saying something in a different way can provide new insights which fellow experts may value.

Experts, given what they know, are unusual in that they are not representative of the vast majority of Audiences. Experts, however, are not entirely unusual. All of us have an urge to fill the gaps in our knowledge. We do not like not knowing something, if the answer is within grasp. The difference between experts and non-experts is one of scale.

> 'People who do not know what they don't know remain blissful in their ignorance, but when people know what they don't know, they are alert to opportunities to find out.'
>
> Source unknown

Experts have solid foundations and generally know what they don't know. That is, they are aware of the gaps in their expert area. They look to fill them as and when they become aware of them. Non-experts have vast realms of ignorance. There is more gap than foundation. A communicator's role is to find a way to bring the Audience into the unknown realm, familiarise them with the World enough to introduce a Big Hook puzzle. The Audience, even with their short acquaintance with the World, need to recognise the Big Hook event as odd, strange or unexpected. It is a gap in their expectations of what is normal in this World. They want an explanation.

Big Hooks can be a Personal Experience

Let us begin by looking at some personal experiences of some successful researchers. These experiences are remembered with hindsight as significant incidents which led to the researchers focusing attention on a particular problem. They were, if you like, the Big Hooks which set the direction for each researcher's academic life. They also reveal a little more about the conditions necessary for a Big Hook puzzle to fully engage an Audience.

As all researchers know, there is no single way of choosing a research topic. For some it is an interesting offer from a respected supervisor; for others the convenient or exciting location of the lab might be a more important criterion. PhD students keen to move onto a new topic once they have their doctorate use networking opportunities at conferences to research their next move. Each person will have a personal, unique account of the influences which attracted them to a particular question in a field of research. Nobel Prize winners may not be your average researchers, but they too during their careers had the same kinds of decisions to make about what to study.

Early in his career the psychologist Daniel Kahneman,[3] was 'casting about for a useful topic of research' when he came across an article in *Scientific American*[4] which inspired him. He was particularly struck by the evidence that the pupils of the eyes are sensitive indicators of mental effort. He simply wanted to know more – much more – about this observation and what opportunities it might offer him if he were to investigate further. Even though he was in a laboratory that was studying hypnosis he decided to turn this idea into a research topic. He found it fruitful. Three decades (and many years of research) later, Kahneman was awarded the 2002 Nobel Prize for Economics 'for having integrated insights from psychological research into economic science, especially concerning human judgment and decision-making under uncertainty'.

Richard Feynman,[5] tells of how he began to delve into the mathematics of spin after an incident in the cafeteria at Cornell University when a student tossed a plate into the air. There was a blue crest on the rim of the plate. Feynman noticed that the plate both spun and wobbled, but the blue crest showed that the spin appeared to be faster than the wobble. The two motions, which were strangely not synchronised, intrigued him enough to seek out the explanation. He turned to Newton's laws to work out the relationship between wobble and spin. He too, many years later, won a Nobel Prize (for Physics), for his work in quantum electrodynamics.

Mathematician Andrew Wiles worked for seven years, much of that time in secret and alone, to work out the proof for Fermat's last theorem. He recalls that his Big Hook for this Herculean task first grabbed him when he was ten-years-old. He found a book at his local library on maths.

'It told a little bit about the history of a problem that someone had resolved 300 years ago, but no-one had ever seen this proof. No-one knew if there was a proof. People ever since had looked for a proof. Here was a problem that I as a ten year old could understand but none of the great mathematicians in the past had been able to resolve. And from that moment, of course, I just tried to solve it myself. Such a challenge, such a beautiful problem.'[6]

In number theory, Fermat's last theorem states that no three positive integers a, b, and c can satisfy the equation:

$$a^n + b^n = c^n$$

for any integer value of n greater than two.

Where n = 2

$$a^2 + b^2 = c^2$$

This is Pythagoras Theorem: the square of the hypotenuse of a right-angled triangle is equal to the sum of the squares of the other two sides. A geometric rule taught in schools.

Each of these giants of research in their everyday World came across a puzzle – something which was surprising and new to them and could not be immediately explained. This is one of the defining attributes of a Big Hook. The other attribute is motivation. The Character in a narrative must not only be curious about the puzzle, he must also care enough to want to pursue the answer. It is literally as if the Big Hook is dragging him out of his chair to search for the solution. How many people who read the *Scientific American* article did not turn their thoughts and actions to a research project associated with pupil size and mental effort? How many other physicists saw the spinning plates in the Cornell cafeteria and were baffled enough to return to their office to do some work on the relevant equations? Why did Andrew Wiles, of all the children who borrowed that particular book from the library, decide that he wanted to spend his working life pursuing the proof for Fermat's last theorem?

Each of these researchers was motivated by a different set of circumstances. Kahneman was in the process of looking for a research topic. His mind was alert to the sense of possibilities and this project provided just such a sense. Feynman was feeling under pressure to deliver important work. He was stifled by it and declared that he wanted to 'play'. The plate spinning exercise allowed him to make his point: he would undertake the project simply because it was 'fun'.

Andrew Wiles' initial motivation was one which many people will recognise. A childhood curiosity which persisted into adulthood, stoked along the way so that it became a defining ambition.

Their motivations were individual, generated through a mixture of circumstance and personality. They were strong enough for each man to devote considerable time, energy and resource in pursuit of their challenges.

When you set up the Big Hook puzzle in a story, the motivation of the Character is something the Audience needs both to understand and to share vicariously.

In fiction, Characters are motivated by human needs – as we have seen in the chapters pertaining to the Audience. In research, the motivation is very often thought to be intellectual curiosity. The curiosity is fed, however, by a set of circumstances. Audiences will engage with scientists' Big Hook puzzles when they share with the scientist some of the significant *reasons to care* about it.

Big Hooks can be an Event

External events – an outbreak of measles, a natural disaster, even a new drug acquiring a licence – are newsworthy. They capture people's attention and concern with the Lure alone. Novelty and surprise are enough. When events continue to make the news however, the narrative needs more than a Lure. The stories need a Big Hook, (referred to by journalists as 'an angle') to sustain Audience interest. Each ongoing event unfolds as many different narratives.

Newspapers, columnists, television programmes, magazine articles and storytellers will focus the attention of the Audience in a particular direction and use it to communicate information which they consider important and interesting. An outbreak of a disease might have a Big Hook puzzle which raises questions around health and sanitation issues, or the loss to business and the local economy. For another Audience the Big Hook might be a question around the virology, bacteriology or history of similar outbreaks.

All are likely to feature Big Hook puzzles which, early on, draw their Audiences' attention to individuals (Characters) directly affected by the outbreak as a way to connect with the Audience, give them reasons to care.

Big Hooks can be a Gritty Challenge

Science researchers, by definition, investigate important and interesting puzzles. They are motivated to explore areas which are not well understood and to solve problems. Science research usually has an inherent Big Hook question which is clear and unambiguous – although it often takes work to make it relevant to the Audience. The Big Hook is the engine which guides and drives researchers to capture their scientific prize: answers to puzzles which lead to new insights and further questions.

More complex challenges, like more complex stories, need to be broken down into smaller components. How to eliminate malaria, how to find the Higgs boson, how to build the tallest building or longest bridge, how to make plants disease-free, are big questions involving dozens of research and project teams and hundreds of people. Each team, each person will be working to understand more about a part of the challenge. Each part of the challenge will have a more precise question and a set of procedures to drill down to the answers. The same principle applies in communication. The more complex it is, the more it needs to be broken down into smaller sections, each with its own Big Hook question. (There is more on this in the *Structure* and *Make your Pitch CRACKLE* chapters.)

The camera crews I worked with at the Open University (OU) in the UK told me that they had applied to work at the OU because they had been interested in the Arts. At first, some had dreaded working on the science programmes because they had anticipated that they would not understand them. What they found was very different. Some of the arts programmes struggled to find a puzzle which would keep the Audience (and them) engaged, whereas science programmes almost always featured an absorbing series of intriguing questions with interesting, and surprisingly comprehensible solutions.

The Big Hook Puzzle You Set is the One You Will Answer – *Which is why it is so important that the right question is asked*

When asked what he would do if he had just one hour left to live, Einstein was reputed to have said that he would spend 55 minutes defining the problem.[7]

An illustration of this point is one of the urban myths which came out of the space race between the Americans and Russians. The astronauts in the 1960s did not have electronic devices like computers and tablets. The Americans set about designing a pen that would write in a gravity-free environment because, so the story goes, their Big Hook puzzle was: 'How to make a pen which will write in space?' The Russians, on the other hand, asked a different question: 'How can we write in space?' They gave their astronauts pencils.

Every storyteller has the option of choosing his or her angle, that is for deciding: 'What is the Big Hook puzzle for my particular narrative at this particular time, with this particular Audience?'

An ultimate skill in creating a narrative is to introduce the Big Hook as a surprising or unexplained event or problem in the World that the Audience recognises, and begins to think about, just before it is overtly pointed out to them. If they become conscious of the

Big Hook question before directed towards it, the Audience feels in control and clever. The storyteller has captured their attention, created a gap in their understanding. She has motivated them to concentrate on following the Plot, and will reward them with the answer to the puzzle – these are the subject of the next chapter.

Questions and Ideas to Help Stimulate Ideas for a Big Hook

Here are some questions to help you create a strong Big Hook for each of your communications. In addition there are some questions to help you think about why an Audience might share your reasons to care about it.

To begin: write down your research question and use this as the starting Big Hook. Alternatively, if your communication is not about your research, use the Change and Affect (refer back to the *Change and Affect* chapter) which you expect your communication to deliver to the Audience. From this extract a Big Hook question which might lead to this Change and Affect.

1. The Big Hook as a research question
    - Write down your research challenge or any other challenge. Ask yourself if you could answer it, what would it allow you to do?
    - If you could do that, what could you do then?
    - Repeat this process until you get to the stupid or meaningless answers.
    - Now look at this new list of challenges and consider whether there is a question among them which has the potential to provide a stronger Big Hook.
    - Alternatively write down your challenge.
    - Keep asking 'Why is this a challenge?' and write down each answer.
    - Why is each of these challenges a problem?
    - As before, a reframed problem may provide a better Big Hook than the one you started with.

2. Why is it important to the Audience?
   - Will your Audience assume the Big Hook question, event, experience, challenge is important?
   - Why will they?
   - What assumptions are you making about what they know?
   - How can you introduce the Big Hook question so that the Audience clearly recognises the importance of the question?

3. What are their reasons to care?

When you are writing for, or presenting to, a non-specialist Audience, the gap between what they know and your World may be large. You may need to reveal reasons why the Audience will care about the important question which drives your narrative.

   - Can you build a sense of importance and mystery into the Big Hook to draw in the Audience? (As in the short opening story at the start of the chapter.)
   - Is there a parallel narrative (discussed in the next chapter) which connects strongly with the Audience, which can introduce your Big Hook? (Obvious examples in medicine are individual stories of patients whose World would potentially be improved if researchers could find the right answers.)
   - Can you provide surprising facts and figures associated with the Big Hook to build its significance for the Audience?

4. What is the prize – your contribution to research?
   - What does your research deliver in new understanding?
   - What is the prize – the results which you will deliver in your communication? Work backwards to identify the Big Hook.
   - What important question do your results provide an answer to?
   - Is it obvious to the Audience that these results answer an important question? If not, work further on the Big Hook to create a stronger one, one that will ensure that the Audience will appreciate and feel satisfied with the prize.

5. What is the prize – your reward?

   If your communication is a talk or a presentation to a general Audience, it may be appropriate to consider whether the satisfaction you get from your work or your motivations, might also provide a Big Hook for the Audience.

   - What has your research delivered that is satisfying to you?
   - How is it satisfying?
   - What important question did it answer for you?
   - What did it reveal about your motivation?

6. The Big Hook as a personal challenge

   - Was there an event during or before your research which captured your attention for being strange, surprising or new and which motivated you?
   - Did you have a 'Big Hook moment' when a potential prize revealed itself?
   - Can you develop either of these into either a Big Hook for your communication or, by referring back to the *Character* chapter, a story to introduce yourself to the Audience?

7

Plot: the perilous journey to solve the puzzle and reach the prize

> 'The secret of being a bore is to tell everything.'
>
> Voltaire[1]

One night, quite late, I had a phone call from Julian Richards. Julian was the presenter of a BBC television series we both worked on. He was excited because, a few minutes before, he had been talking to a former colleague and fellow archaeologist who had just returned from visiting a building site in Bristol. He had been called there by developers who had come across what looked like a stone coffin, a sarcophagus, bang in the middle of a car park they were digging up. The developers wanted the archaeologist's help to remove it.

Julian thought we should send a film crew to follow the action first thing in the morning. He did not have much to go on, except the Big Hook of a sarcophagus. It was a huge mystery which screamed out questions. What was it doing under a parking lot in Bristol? Who, if anyone, was buried inside? What else might there be inside, or buried nearby? He had a strong hunch that this discovery might develop into a potential story for our archaeology series

Meet the Ancestors. We decided that the cost of filming the removal of the sarcophagus from the car park was worth the gamble. If nothing was found inside, we would almost certainly have no story. Yet if the stone coffin contained a skeleton or perhaps some burial objects, we would have many more questions, more puzzles, but also lots more avenues for investigation. Each one of these avenues had the potential to provide clues to help solve the mystery around the sarcophagus. This, for *Meet the Ancestors*, was what we required for our Plot.

The Plot is an account of the events or episodes which follow when a Character decides he wants something badly enough to make efforts – sometimes strenuous, sometimes risky – to reach his goal. It is about the actions a Character takes to overcome all the Obstacles blocking his way; to triumph or fail in his challenge.

The majority of narrative Plots within science research are like each edition of *Meet the Ancestors* – straightforward quests. Science itself is a quest; a searching journey by the curious, a hunt to find things and find things out. Most science, however humdrum and ordinary it may seem to the researcher at times, has qualities of strangeness which are interesting to an Audience. Research, by its very nature, has a natural story structure so it has the potential to be made dramatic and absorbing for an Audience.

The communication task is to elevate what at first sight might seem an ordinary account into one that is as engaging as it can legitimately be, both for the Audience and for the creator. It means creating a Plot.

In this chapter I shall refer briefly to three types of Plot which most commonly apply to communication made by scientists. They share the same basic components and follow a similar pattern. I shall use the simplest of narrative examples to show how a Plot is constructed, and how it can be improved to make the story better. This does not mean altering the facts, it means choosing which episodes to include and making them work for the story. In other words, using every opportunity to make the story meaningful for the Audience; giving it *crackle and fizz*.

The Different Kinds of Plot

The nature of a Plot defines the kind of story that is being told. Science is a human endeavour and since stories are about the human experience, examples of every kind of Plot can be found in the encyclopaedia of science tales. Yet, surprisingly, there are remarkably few Plot types. Indeed, an argument has been made that every conceivable story, from the Greek tragedies to soap operas, can be classified by seven basic narrative Plots.[2] I would go further to suggest that there are just three Plot categories which cover most narratives in science research: adventure, survival and fellowship.

Adventure

Adventure stories are narratives about exploration. They are journeys into the unknown. They can be large expeditions, for example, to understand climatic conditions in Antarctica, to map the ocean floor, or they might be a piece of work which simply pushes the boundaries of what is known or possible. For example, isolating a new protein or creating a novel smart material. Adventure is a mindset; there is not necessarily a practical or profitable goal – although there often is. (Blue-sky research is an adventure – an adventure for the insatiably curious.)

Characters in adventure narratives are motivated by an enthusiasm for discovery. Adventure stories guarantee that their Characters will experience periods of risk, danger and excitement, and hazards and thrills.

Survival

Survival stories in science can be adventures too, but the stakes are higher: lives are in jeopardy. Medical research and some engineering projects are typically about survival: improving the lot of humans and other species.

Characters in survival stories, unlike adventure stories, have not usually chosen their role; the role has been thrust upon them by their situation.

Fellowship

Stories about fellowship are those which scientists share with each other in social settings – over drinks or meals at a conference, for example. They are stories about colleagues or themselves which confirm the values and behaviours which bind scientists together as part of wider community. As the biologist and Nobel laureate Peter Medawar noted in his *Advice to a Young Scientist,* there is no such thing as a typical scientist. They are as different from each other as any group of people.

> 'Among scientists are collectors, classifiers, and compulsive tidiers-up; many are detectives by temperament and many are explorers; some are artists and others artisans.'[3]

Each branch of science has a different history, different rules of engagement, and different physical processes and procedures. Narratives reveal these differences. To truly understand an area of research, like any aspect of life, one needs to know its fellowship stories.

Although every Plot is, of course, different in the details, almost all Plots, including the adventure, survival and fellowship Plots, follow the same basic pattern. It is a pattern with which we become familiar very early in life. Fairy tales, read or told to children in every culture introduce the story framework and expose them to this Plot pattern. Jack and the Beanstalk – the story of a poor boy who goes on an adventure and becomes rich. Cinderella – the tale of a lonely victimised girl who, with the help of a benefactor, survives her circumstances to find wealth and romance with a Prince. Androcles and the Lion – who discover, from Aesop's classic fellowship fable, that kindness to others brings rewards.

So what is this pattern? It consists of five constituent parts: **C**haracter, **O**bstacles, including a **F**inal Obstacle, **A** Turning Point and a **D**enouement (**COFAD**).

The *Character.* They want something. In science communication they are frequently the presenter, author or the Audience themselves, but Characters can also be an animal, a vegetable or indeed a mineral.

The *Obstacles.* These are in the way of the Character getting what she wants. They can be physical or mental, real or imaginary. They require creativity, effort and action to quash, make irrelevant, use or work around.

The *Final Obstacle* is the toughest of them all. It tests the Character to her limits.

A Turning Point is the crossroad. Procrastination is over, decision time is here. The Character is forced to choose which way to go. It is a moment of drama which creates tension in the Audience just before the Denouement.

The *Denouement* is where we discover whether or not the Character achieves her goal. It ties up the loose ends in the story and it delivers a satisfying outcome for the Audience.

Let us begin by illustrating the principles of COFAD with a simple quest story, the kind of anecdote any proud parent might relate to a relative. It may not, at first sight, seem relevant to communications about science or research, where the quest appears to be quite different, but bear with me because I shall show how the simple story illustrates the Plot principles very directly. I shall then show how creating a narrative about a research project requires thinking about the project in a way which can exploit these principles.

The simple and direct quest in summary:

John is a toddler. He wants the mobile phone which has been deliberately placed out of his reach. He eventually finds an implement, a backscratcher, and uses it to shift the phone into a position where it will drop onto the floor and be his.

Character in Search of a Prize

In any Plot there is a Character in search of a prize. In research the prize may be a cure for pancreatic cancer, a way to make solar cells give up more of their energy or smart materials which can improve technology applications. In *Meet the Ancestors* the prize was a significant story, and an insight into the life of a Briton of a bygone age. John's quest is much simpler. He wants the mobile phone which is out of reach.

In the *Character* chapter, we looked at how to ensure that the Audience cares about the Character, so we shall not dwell on this here. The important point to make again about the Character is that the Audience needs to understand, or during the story to find out, why the Character wants his prize. His motivation may not be obvious.

For the purposes of Plot and this example please assume that you share an adoring parent's fascination with John and his development.

Obstacles

The main Obstacle to John getting hold of the mobile is that it is out of his reach. The scale and nature of the Obstacle are clear, and it is not hard to see that it presents a challenge for a toddler. It is an Obstacle which does not need further explanation.

In our *Meet the Ancestors* programme the nature and scale of the Obstacles to finding out about an individual ancestor were more complex and not apparent at the start of each project. The first Obstacle in our quest however, was clear: getting the burial out of the ground. This too, does not require further exposition.

In a narrative about research the Audience needs to understand, as clearly as they understand that the mobile phone is out of reach of the toddler and the sarcophagus is still in the ground, why and how the Obstacle is creating a barrier in between you and your goal. In a research quest, this may not always be apparent to an Audience.

While in the instance of the first Obstacle in *Meet the Ancestors* it was blindingly obvious why it was a block to getting our goal, thereafter it was very different. Once the bones were excavated, the Obstacle which remained was so huge that it was difficult to know how to define it simply. It could not be understood in the same way that an object beyond arm's reach can be visualised. There was a vast information gap to bridge from the ancient skeleton itself to a consequential history and image of an individual who lived thousands of years ago. The strength of our Plot would require that our Obstacles be made apparent, clarified.

In research there is a similar issue. Research goals are relatively clear in their aims, for example improving human health and ensuring the world population has supplies of fresh water and food. If the goals are very ambitious the Obstacles in the way become larger and more evident as time goes on. Obstacles are not usually manageable, definable entities which declare themselves upfront or in clear quests. There are no maps with the Obstacles nicely signposted. I have heard scientists describe research as 'like walking in a fog'. You stumble around knocking into things, walking into dead ends, hoping that, along the way, you find something to suggest that you are on track, taking you closer to your goal.

On a quest, whether it is a research project or creating a narrative of it, the requirement is to be open to ideas, to look around, to see what emerges that might be of use in breaking down some of the Obstacle and getting you closer to your goal. We often spent several days, and occasionally longer, filming the retrieval of the bones for *Meet the Ancestors*. Such excavations were thorough, painstaking and slow; they were only interesting when there were some surprises: the discovery of a burial article, some loose teeth or a tiny piece of preserved clothing.

When it came to putting the documentary together there was usually little from the excavation worth including in the narrative. Yet, knowing this, we still always filmed the entire excavation. The reason we did so was because at the start of the process we had little or no idea what parts of the excavation would turn out to give us interesting and surprising ways to get closer to our goal – constructing a history and an image for one of our ancestors. The excavation was our first area for searching for things which might reveal a way to make some of the Obstacles accessible and help us on our way to discovering something about our ancestor.

In *Meet the Ancestors*, we had to use information from many sources: the artefacts we recovered, facts we could retrieve from similar burials elsewhere or from the history of the location itself. Taken together the information gave us opportunities to pursue a line of questioning which led to separate mini-narratives or subplots. For example, we asked ourselves what the teeth from the sarcophagus could tell us about our person. We were able to use the teeth as the means of creating a subplot. We could use this subplot to solve the puzzle of where the person grew up.

Subplots and mini-narratives are episodes which are driven by a more precise, narrowly focused question than the Big Hook. These questions are relevant to the overall challenge and take us nearer to the goal. Taken as a group, the subplots are episodes which drive the Plot forward. Each subplot uses the COFAD structure.

To summarise: the Audience must understand why and how each Obstacle is a block to the desired goal.

John's attempts to overcome the Obstacle in the way of getting his mobile phone are his actions driven by his thoughts. What we look for in good Plots are clever, surprising ways the Character uses to overcome Obstacles and get closer to his goal. The action described in the narrative could well be along these lines:

John's first unsuccessful attempt to get the phone is to scream and stamp his feet in frustration. It cuts no ice with the adults around him – even the dog ignores him. His second attempt is to sidle up to an adult and pull them towards the table, smiling and pointing upwards. This does not work either.

Later, left alone in the room, he slides a stool towards the table. Standing on it he can see the mobile, but is still unable to reach it. Banging the table with his fist and shaking the table leg does not shift the phone. He looks around and spies, on the sofa, a backscratcher which he has seen others use. He picks it up and has an idea.

This short piece illustrates two attempts, both of which fail, to get the phone.

It draws attention to John's challenge, and involves the reader in wondering what the toddler is going to do to get his prize. Clearly, because it is being told as a story, we expect him to either succeed or fail in an interesting way. So far he has done nothing unusual for a toddler. (Notice that in this section it is noted that 'even the dog ignores it'. This is part of the set-up. It was added for a reason which will become clear later.)

If the Obstacle in your research project is a common one and is not essential for understanding the research (for example, securing funding or getting access to a piece of equipment which is only available at certain times) and unless you have found an unusual way of dealing with the Obstacle which might be of interest to the Audience, then do not include it in the narrative. If the Obstacle is essential to understanding the narrative but is not particularly interesting, include it but do not dwell on it.

On *Meet the Ancestors*, we did not include much of the excavation in the final telling of the story for this very reason. It occasionally upset the team who felt that they had put in an enormous amount of effort – getting up early, staying away from home for weeks – with nothing to show for it. Sometimes the excavations revealed so little that we did not have enough material to begin to ask questions of it or to find out more about it. We had to abandon our quest. That, sadly, like the nature of research, is just how it is. It is not, however, relevant to narrative, where Audiences expect to find something of interest and for something to happen.

When John finds the backscratcher, the Audience sees that he has found an ingenious and unexpected way of trying to get his prize. If toddlers around the world used backscratchers as a standard method of obtaining objects placed out of range, the action would be insignificant and the story would not be worth telling; but toddlers do not, so the Audience is mildly intrigued to see what happens next.

He stands on the stool holding the backscratcher in both hands and tries to hook the backscratcher over the phone. It is a tricky task for the toddler, as the phone keeps slipping out of contact with the tool. He manages to get traction with the phone, momentarily, and it slides at an angle a little way across the table. He tries again and it spins around. Another attempt and it slides off in a slightly different direction...

Once the Audience is engrossed in the puzzle and sees the possibility of a solution, they are willing to accept detail. Indeed in many stories they *want* the detail. It helps them (as well as the Character) to get closer to understanding how the puzzle will be solved.

Researchers frequently suggest that they are unable to give the full particulars of their science because the Audience is not interested and will be unable to follow it. Not so. Once they understand the challenge and become party to solving it, they can find the detail satisfying and rewarding. The Audience does not mind concentrating on details and devoting effort to a task if they have invested emotionally in solving the puzzle.

If at the beginning of a talk a researcher begins, as they often do, with the details of how something works, it is very unlikely the Audience will commit to listening attentively. If, however, the same details are given within the Plot, then once the Audience understands how the details might contribute to overcoming a particular Obstacle in the way of the prize, the Audience are likely to be much more receptive.

Back to John's attempts to get the phone. He has virtually overcome the Obstacle that he had in view when he started, but he is about to discover that his path to the mobile is not quite as clear as he thought...

Final Obstacle

In most stories, there will be more than one single Obstacle in the way of the prize. In research projects, not all the Obstacles are known at the beginning. They often emerge as the work progresses. Obstacles become significant to a narrative when they demand the attention of the Character because dealing with them provides the means to get closer to the goal. Each Obstacle needs to present new difficulties for the Character and different problems to solve, preferably under increasing pressure.

A good narrative feels like a journey through an Obstacle course rather than a hurdling race. Even something as simple as television quiz show, for example *Who Wants To Be A Millionaire?*, follows the COFAD Plot pattern. The questions (that is, Obstacles) get more difficult and the pressure mounts as the prize of becoming a millionaire comes closer.

The Final Obstacle is the one which presents the greatest challenge for the Character. It is the time where the Character's true nature is revealed because it presents him with a difficult choice. In *Who Wants To Be A Millionaire?*, the contestants are required to choose whether to accept the money they have won so far and leave the game early, or whether to progress to the next question which, if they get it wrong, means they lose almost everything. If they get it right, however, their winnings will be doubled. It tests a person's capacity for self-confidence, risk, courage or greed.

All Obstacles demand that the Character makes choices about what to do. Often these choices are not particularly testing. For example, if you are running late for work, you may have a choice of driving the certain long route to work which would get you there – possibly a minute late – or gambling on the shortcut which, if there is a truck in front of you, will certainly take longer. Being a few minutes late is not usually important, but if your life depended on catching a plane that would not be there if you were late, your choice would be vitally important. Your choice would reveal something about you. (My father always advises, 'if you are in a hurry never take a shortcut you have not used before'.)

If John gets the phone to the edge of the table and then collects it as it falls to the floor, the narrative is not as interesting as when there is a Final Obstacle. The Final Obstacle for John is the adult who took the phone away from him:

As the phone reaches the point where a final pull will bring it within grasp, an adult walks into the room. Not just any adult but the adult who had put the phone out of reach quite deliberately. The toddler and the adult look at each other.

A Turning Point is upon us.

A Turning Point

John sees the adult and in his mental conflict we have A Turning Point, the point of no return, the story's crisis and the critical point in the story where the toddler is in most jeopardy (in the narrative). He has an important choice to make; one which will affect his success and, in some ways, define him. He needs to make his decision: to continue and, if he is quick enough, get the phone as it falls onto the floor, or to give up and fail, on this occasion, to get his prize. Of course, there may be other consequences which he is unaware of. He may win the phone, but lose out because the adult chooses to crush his creativity. He may decide to give up, but nevertheless be rewarded for his ingenuity in another way.

The Audience recognises John's dilemma. He is at his most vulnerable and exposed. They are curious about what he will do. Will he be bold and brave or play it safe and give up the attempt? They are rapidly running through the various options open to him and guessing which one he will choose, thinking what they would do in the same situation. There is suspense at A Turning Point.

> 'In constructing the plot, the poet should place the scene, as far as possible, before his eyes. In this way, seeing everything with utmost vividness, as if he were a spectator of the action, he will be most unlikely to overlook inconsistencies.'
>
> Aristotle[4]

The suspense can be extended by following the advice of Aristotle. He suggested that the best way of maintaining the tension in a Plot is to slow the action by describing the critical scene in vivid detail. By doing this you build suspense and force the Audience to wait before you satisfy their curiosity with the Denouement. This is done very deliberately in television quiz shows where people are called upon to make their choice. The hosts of the programmes do not want the contestants to make their decision quickly. They want to milk the situation for all of its jeopardy and tension, and thus allow the Audience – in what they hope is a heightened emotional state – to enjoy the uncertainty of the moment at A Turning Point.

Denouement without a Twist

At the end of a narrative is the Denouement. This is the part which completes the story. It should leave the Audience feeling satisfied that the final act is consistent with the rest of the narrative. They expect the end to inform them whether the Character succeeded in whole, or in part, in getting his prize or whether he failed. The Denouement must be congruous with the Big Hook puzzle, it must be truthful about the Character's motivations in overcoming the Obstacles, and the choices he has made. It delivers the Change and Affect (discussed in Chapter 2).

The Denouement for our simple quest could be this:

> *The toddler makes his decision: he grabs at the phone. It falls; he steps off the stool, picks it up and runs as fast as he can out of the room, his mission accomplished.*

However, stories which are the most memorable contain a surprise at the end, a twist in the tale.

Denouement with a Twist

Narrative twists can be enough to make a story worth telling. It is how jokes work. They lead the Audience to a certain expectation and then surprise them with a totally unexpected, yet satisfying, Denouement. The actions which the Character takes at the crisis point delivers something which he and the Audience were not expecting. So the phone might fall towards the floor and the Audience might be expecting it to shatter, or to be plucked from the infant by the adult. They might be expecting the adult to be amused by the child's ingenuity or annoyed by his disobedience.

All of these options will be running through the Audience's mind. The twist has to offer something the Audience had *not* considered.

> *The toddler makes his decision: he tugs wildly at the phone with the backscratcher and it slides off at an angle to the very edge of the table. As it falls, as if in slow motion, the dog comes to life. The phone hits him on the nose before bouncing on the floor. He catches it in his mouth – and he is away. The toddler and adult set off in pursuit.*

The twist must be original if it is to work. If we are aware that others have used it before, then it is unlikely to have the power to surprise us. We have probably all seen movies where a Character is apparently shot dead, only to reappear to take a critical role at the movie's Denouement. That particular Plot twist no longer has the power to astonish us.

The Plot twist should not feel like an add-on. Stories do not introduce new significant Characters at their end, nor do they bring in anything else that is new to the World. If you introduce a Character – like the dog – or an object, however fleetingly, at the beginning of a narrative, you can bring them back at the end and the Audience will accept them because they have been shown to be consistent with the situation which has already been set up.

Once the Plot has concluded with the prize won or lost, the Audience expects that all their unanswered questions will be resolved and all the loose ends tied up. Anything introduced in the World with a small hook, a query which the Audience has invested thought into, needs to be answered. You do not want an Audience leaving a story about an important advance in cell division, wondering what happened to the six petri dishes which you described as having been taken from the fridge and left, accidentally, on the bench overnight. Those amongst your Audience who are paying close attention will want to know what happened to them and why they were ever mentioned.

And finally:

If You've Got a Wooden Leg: Wave it Around

I once worked with a public relations officer, a former journalist, who was fond of saying, in jest, 'never let the facts get in the way of a good story'. It was usually when he was exasperated that the science stories we were promoting did not fall into the best Plot pattern to grip an Audience. I learned to take another view, one which I acquired from one of my former bosses. He always used to say, 'if you've got a wooden leg: wave it around'. It took me a long time to understand what he meant. After a few years, I realised that it was the direct opposite of 'never letting the facts get in the way of a good story'. It was conversely, '*always* let the facts get in the way of a good story'. Not only let them get in the way, but make them a feature. This is relevant to research because sometimes certain pieces of evidence do not fit with the theory. In research this is something which cannot be ignored, but it does not have to make the narrative less interesting for the Audience. Let me illustrate this with an example from my filming days.

I was making a film on a farm about a new potato harvester. The weather was cloudy and soon it began to rain. Because I was inexperienced I carried on trying to ignore it. I was optimistic because I knew that rain drops are often difficult to see on camera. It became clear, however, that by the afternoon the presenter, standing out in the field without an umbrella, was completely drenched. Her hair was soaking wet and clung to her face. Water ran down her forehead and streamed off her nose. All the while she bravely delivered her lines.

Later, in the cutting room, it was clear that the rain was a big Obstacle to our telling of the story. I had chosen to ignore it, hoping very naïvely that no-one would notice!

What I learned is that Obstacles can be made part of the Plot. Instead of ignoring the rain, I should have used it and drawn the Audience's attention to the weather, making it a feature of the story — shown it as an Obstacle which must be dealt with.

So, on another occasion when a colleague rang to tell me that our presenter had broken his leg getting off a plane, but not to worry, she would film him in close-up, so the Audience would never know, we discussed a different option. Make it part – albeit a small part – of the film. The Audience may be intrigued by the unexpectedness of this extra episode, and it will make the rest of the filming easier – particularly for the presenter who does not have to disguise the fact that his leg in plaster!

When facts emerge which challenge or overturn the very assumptions on which the project is based, the narrative must change to take account of the new reality. It is what distinguishes non-fiction from fiction. The narrative needs to unravel in a way which both meets the Audience expectations and is truthful. This depends not only on the substance and content of the story, but how it is put together and how it is structured. Structure is the subject of the next chapter.

Questions and Ideas to Help Stimulate Ideas for Constructing your Plot

Here are some questions which are designed to stimulate you to think about all the ingredients of Plot. If you have, as many researchers do, a number of mini-plots we shall consider how they can go together to maintain interest and create suspense and tension in the *Structure* chapter. Here we are looking at the single linear Plot. It is worthwhile, however, to look at all the mini-plots as linear Plots. It provides you with lots of options for the COFAD ingredients which you can draw on for the story you want to tell. As always, do not censor your ideas until you have noted them and are considering the Audience for your narrative Plot.

Constructing the COFAD for a single Plot

Take a blank sheet of A4 (or larger) paper and place it in front of you in a landscape format. Draw a line along the bottom to represent a timeline of the project – the Plot line.

1. Denouement

   At the far right end of the Plot line begin by noting (or drawing) the Denouement. Write down the conclusions of the project, or if it is a work in progress report, the take home messages of the communication.

- What is the end-point?
- Is it a definable goal or is it a step towards a goal?
- What is its significance?
- What is the Change and Affect you are delivering to the Audience with this Plot? (There is more on how to extract the most from your Denouement in the *Change and Affect* chapter.)

2. Character

Who is the Character in this narrative Plot?

- Is it you or someone else?
- Could a version of this narrative be made with a non-human Character, for example a fruit fly, a cell or a mountain? Again, even though in reality it may be absurd to believe they have 'a goal', in narrative terms they can.
- On the left hand end of the timeline, place the name of the Character and their goal. For the remaining elements of the COFAD, we shall assume that the Character is you.

3. Obstacles

These can be known unknowns. They can be lack of resources or individual people – competitors or others.

Draw a relevant shape to represent each of the Obstacles which were known at the beginning of the project on the timeline. (Include obtaining funding, finding a lab or anything else which may seem peripheral – it could be useful.) Draw them along the timeline when they became known.

- Against each Obstacle note the new ideas or information (or people) you sought out and used to help you overcome it to get closer to your goal. Summarise them.
- Was there anything ingenious, innovative, surprising or unexpected in how you did this for any of the Obstacles?
- Did overcoming any single Obstacle create a new one? Draw this on the timeline closer to the goal.

Looking at the timeline with the Obstacles in place, were there any other significant events which took you closer to your goal?

- What were they?
- How did they take you nearer to the goal?
- Do they reveal another Obstacle which you had not considered as an Obstacle? If so, add it to the timeline.

4. Final Obstacle
   Is there a Final Obstacle? Something which was particularly challenging towards the end of the project and blocking progress but which was solved? Don't worry if there is none. However, do look at the questions in Section 6, below, just in case you are able to adjust your end-point to create a Final Obstacle and a more satisfying Denouement.

5. A Turning Point
   - When did you realise that you had reached your goal? (Or end-point in the narrative if it is a progress report.)
   - Was there a single incident or event which gave you insight or confirmed your insight? Sometimes these can come from 'left-of-field', that is by accident or from a piece of serendipity.
   - Was there a point when you needed to make difficult choices or decisions? Note these down next to the Obstacles which they apply to.
   - Is there a decision to be made at the point of almost reaching the goal, or approaching a significant point in the project? Would this decision be testing or difficult? Would it surprise the Audience? Could this be A Turning Point?

6. Denouement revisited
   You should now have the project laid out so that it shows, in detail, the content available to use to construct your narrative Plot. Ask yourself the following questions:

- Are you happy that the Obstacles work towards the stated goal, answers to the Big Hook question?
- Are all the questions raised in the narrative consistent with what is delivered in the Denouement?
- Is there a significant, different Final Obstacle which would give you an alternative final goal, a different Denouement? One perhaps which has a stronger turning point – a moment of crisis — to make the narrative more memorable. For example, your Denouement might, at the start of your thinking, be a final experiment which supports an idea or hypothesis you have. After you have gone through this exercise, another alternative Denouement might be relevant: a decision to move to a different group, a struggle to gain support, an offer of a job in a new sector — any number of consequences which may occur.

Remember, you can choose where to begin and to end your narrative as long as you remain true to the overall outcome of events. It would, of course, be totally unacceptable to end a story of 'an important paper published in a prestigious magazine' at the point of publication if you knew that shortly afterwards the paper was shown to be fraudulent.

8

Structure: making the whole more than the sum of its parts

'A lecturer should exert his utmost effort to gain completely the mind and attention of his audience, and irresistibly make them join in his ideas to the end of the subject......

A flame should be lighted at the commencement and kept alive with unremitting splendour to the end.'

Michael Faraday[1]

There is an illustrated children's book entitled *Zoom*[2] by Istvan Banyai. Not a single word lies within the covers of this book, yet it has a strong narrative. Each page is a single picture which is connected in an obvious way to the one before it, and the one after it. Only when seen as a whole is the presence of a story evident.

I came across *Zoom* when I took part in a creative leadership exercise. Thirty-two people were each given a single colour picture, a page of the book, and told not to show it to anyone else. The group task was to arrange the pictures on the floor face down in an order which made sense, and to do so as quickly as we could.

Each of us needed to find people who had similar pictures to our own. We formed small groups. One of the groups had an aeroplane in their picture, another a ship, others a cockerel. Some had a sea view, others a desert view. There were children and adults from different parts of the world. Within some groups it became obvious that the scale of the common objects was different in each picture. After a while someone decided to write down a list of the groups, and people gathered round to add information. We were looking for some connecting idea that would take us from our individual World to the World on the previous and next pages of the sequence. One or two people suggested that the key to understanding the puzzle was to look at it as a zoom in to, or zoom out from something. This was the turning point. Soon groups were forming within the subject themes. What turned out to be the first group lined up at the front. Their first picture was a tiny, distant planet Earth. By the fourth zoom in, the Earth was being circled by a plane. By the ninth picture the plane was no longer visible but we were looking down on a group of local people standing on a tropical beach. By the eleventh picture the postman on the beach had delivered a postcard with a USA stamp mark. By the sixteenth picture we were looking at a completely different scene, a detail of the stamp: a scene on board a cruise ship which itself was a picture on a poster mounted on the side of a coach driving through a city. From there a series of similar zooms took us from the front cover of a magazine being read beside a swimming pool into a children's toy world with a cockerel. The final picture: a strange big close up of the cockerel's comb.

In this exercise there were only two ways to arrange the pictures to make a coherent narrative. Either the narrative started out in space and zoomed in or, alternatively, it began with a child's toy and ended in outer space. I had, until writing this chapter, always assumed *Zoom* was a zoom in. I took the meaning to be a demonstration of the connections which link us to others from different cultures and continents across the planet. Individually, each picture opened up the World of the previous picture by adding more detail, providing, in 31 steps, a narrative which delivered a sense of the interconnectedness of people on our shared planet.

I only discovered the author's intended Structure when I came to find the reference for it when writing this book. Istvan Banyai did not structure it as a zoom in, but as a zoom out. His Structure was prescribed by his purpose which was to impress upon the Audience that things are not always what they seem. His Structure made the book an adventure. Each picture was a challenge for the readers, a puzzle to work out. The Structure, as I assumed it to be, was a pleasant enough journey but there was no mystery; no sense of discovery.

Most narratives, indeed most communications, have many more than two ways in which they can be put together. There is no single right way to tell a story or communicate an idea. There are wrong ways and better ways. As with a jigsaw, you can complete the picture by starting at the edges, at the corners, or by tackling the big easily identifiable objects in the middle. Whatever tactics you follow – where you start and how you continue – the order the pieces are put together does not influence the final image. It is hugely different for communication. The final take home message of a communication is entirely dependent on how the content is put together. Its Structure and delivery are as important as the individual components which make up the content. They both determine how the Audience engage with, and make meaning from, the communication.

Having the Basic Content

In this chapter we shall assume that, at your fingertips, you have your content choices. You are familiar with your ingredients: the World and the possibilities for the Big Hook; you understand your purpose – the Change and Affect you want your narrative communication to produce; you have thought about your motivations and those of any other Characters; and you have identified the Plot, including the significant Obstacles in the way of the goal and your ways around them.

Having the Occasion

The task is to put the basic content together in such a way that the Change and Affect does the job you would like it to with the Audience you have in mind. If you put it together carelessly, it can impress the Audience in ways you might not want, as this example from the World of politics demonstrates.

When Ross Perot stood as a US Presidential Candidate in 1992, his campaign manager, Frank Luntz, organised a focus group to assess the appeal of three television advertisements: a biography, a speech and testimonials. When the advertisements were played out with the biography first and the testimonial last, the opinions of the focus group mirrored the general public opinion polls of Perot. When, however, they were played (by accident) with the testimonial first and the biography last, the focus group liked Perot less than the general population.[3] Structure matters because it affects how Audiences react to the material; it influences how they think. This applies to all communication, whether it is a poster presentation or an expression of a point of view at a meeting.

Having a Go

In this chapter we shall look at how to Structure communication using examples which have one or more of the following broad purposes: to inform, to educate or to entertain.

When the material, the Audience or the experience of the format is new to you, physically putting the pieces and ideas together, assessing what works, trying it again with adjustments and continuing in that improvement loop is time consuming but, alas, absolutely necessary for excellent results. As a remark often attributed to Mark Twain goes:

> 'If you want me to give you a two-hour presentation I am ready today. If you want only a five-minute speech, it will take me two weeks to prepare.'

The purpose of creating a Structure is to maintain the impetus for Audience engagement throughout the whole communication. The narrative rules, which we have met in previous chapters, apply. In summary they are to:

Give the Audience reasons to care about the Character and their World.

Create gaps in knowledge: mystery arouses the curiosity of Audiences. They do not like not knowing something when it could be within grasp.

Build suspense: make the Audience anticipate the answers, wait with concern for them.

Create crackle and fizz: surprise the Audience, keep them emotionally and intellectually engaged throughout.

In the next chapter we shall look at how to prepare short persuasive arguments to convince an Audience of their merits. We shall illustrate how to Structure a persuasive pitch with the *CRACKLE* formula, which is particularly useful when applying for grants or other rationed resources. In this chapter my intention is to illustrate, through examples, a number of different narrative Structures that are commonly used in science communication.

A Single Narrative Thread

Tomorrow's World was a magazine programme which featured new developments in the worlds of science, medicine and technology. It ran for nearly 40 years on BBC television and attracted, in its heyday, as many as ten million viewers (approximately 20% of the UK population) a week. I worked on it during the 1980s. Every other week, each of the producers would be responsible for producing a single story about some new development or application of science. We went to great lengths to find items which had important problems with interesting Plots, that is, problems which also required ingenuity to solve and, in the process, delivered surprises.

Each individual story was usually a quest. It had to be told in anything between three and six minutes. There used to be a way of telling the story which every newcomer learned when they arrived. First, you find something surprising, unusual and interesting to catch attention, set a scene, and then fill in enough detail for the

Audience to feel they understand the World they are in. Second, you introduce the Big Hook; that is, show something which draws attention to a particular puzzle or problem within the World which the Audience sees as a difficulty in need of a solution. Third, you show how many have tried – and failed – to solve the puzzle in the past, and finally you introduce the 'But now…' phrase, which heralds the new development together with an explanation of how the ingenious application of science, engineering and/or technology managed to overcome the problem and leave the World a better place.

My four minute film about the new lifts installed in the Eiffel tower was a typical quest; its narrative Structure, in summary, was as follows:

Narrative element	Purpose	Eiffel tower example
Lure	Grab attention.	Used the World of Paris and the Eiffel tower as the backdrop for the presenter. The Eiffel tower is the most visited paid-entry site in the world.
World	Focus on the main attraction.	A few sentences on the history of the tower, its construction and the spectacular view from the top.
Character	Show who is involved.	Queuing tourists from around the world. In winter, however, they are turned away.
Big Hook	Present the challenge, reveal the goal.	How can the lifts be kept open in bad weather and strong winds?
Plot	Reveal the actions taken to overcome the problem and achieve the goal.	Outlined the old idea to make the lifts safe by encasing them in a metal shaft – it was rejected for aesthetic reasons. Introduced new idea: keep the lift cables steady with an ingenious series of barriers which rise just before the lift reaches them, and close just after it passes. Explained how moving magnets on the lift, passing over metal strips on the shaft, were able to generate a current which created a signal to control the opening or closing of the barriers, thus protecting the lift cables in strong winds.

(Continued)

(Continued)

Narrative element	Purpose	Eiffel tower example
Change and Affect	Deliver Denouement – answers to the Big Hook problem, and creates a mood of optimism.	Public happier because they can now visit top of tower all year round. Officials at Eiffel tower happy because of increased profitability.

This is the typical Structure of a narrative quest. On *Tomorrow's World* it was a Structure which helped us to engage millions of people – most with little formal education in science – in some of the scientific, technological and medical issues of the day.

Let us look in more detail at this quest Structure to draw out some general points about designing: the *beginning*, the *middle* and the *end*.

The End

It may seem perverse, but if your communication is a narrative quest, once you have all your material, the place to focus thought is not at the start but at the *end* of your account.

The purpose of narrative is to provide a frame for the Audience to view and interpret ideas. In the Eiffel tower example, my purpose was clear. I wanted to reveal some brand new lifts in the Eiffel tower which worked in an ingenious way. This was my Change. The Affect I wanted to create for the Audience was a sense of privilege associated with having an insight into the workings of one of the world's most impressive monuments. I wanted them to know something which would make any future visit to the attraction more fulfilling.

Being clear about your purpose is critical. If you are asked to report results at a large conference or share them with coworkers, you are likely to have a clear idea of where your communication ends. If your presentation could be thought of, in metaphorical terms, as the narrative equivalent of 'slaying a dragon', then what state is the dragon in, and what state are you – its slayer – in, at the end of your presentation? The answers represent your Change and Affect.

If you are clear in your mind about these, it will help you set the boundaries for the frame of your communication and give focus in the building of the overall Structure.

The Beginning

This can be the part of the communication which takes the most thought. It is where you have the best chance, sometimes your only chance, of connecting with an Audience. The Lure, as we saw in the *Lure* chapter is the only ingredient which has a single place in a Structure; at the very beginning. It gains the attention of the Audience to draw them in to the communication.

In my Eiffel tower story, for a mass television Audience, my Lure was Paris and the Eiffel tower itself, a straightforward beginning which also set the scene, simply and quickly.

Audiences, particularly professional scientists, who have not signed up primarily for entertainment, often prefer the Lure to summarise the story so they can decide whether or not to give it their attention. This is the principle by which news operates. The beginning of a news report is where most of the writing effort is directed.

The Inverted Pyramid

News items generally follow the Structure known as the 'inverted pyramid'. It is a Structure learned by journalists which is evident every time you pick up a newspaper, and watch or listen to the news. The headline, which is the Lure, also has the guts of the story.

The headline is both Lure and story essence

The first paragraph has the who? when? where? how? And why?

The rest fills in the less important details

News editors who have traditionally decided what the news is, expect that readers and listeners are keen for the essential unadorned information. In the case of a breaking news story they know that *surprise* is their main way of grabbing their Audience's attention. By presenting the bare essentials they give the Audience enough facts to make a decision about how much attention they pay to the entire story. The bare essentials in journalism are the: *who? when? where? how? and why?*

In a continuing story, there is no longer surprise, but an appetite to *fill the gaps* in knowledge. Feature articles, as we discovered in the *Big Hook* chapter, generally follow a narrative pattern similar to the quest which builds to reveal the Change and Affect.

Formal science papers whose purpose is also to inform usually begin with a paragraph which explains, in summary, the essence of the paper. They, like succinct news stories, are not easy to write as they need to convey what is important, as well as provide a sense of proportion. They need to leave out what is not essential to the Audience being able to understand the Change and Affect.

When scientists at CERN gave notice of a meeting to report their discovery of the Higgs boson, using the Large Hadron Collider in Switzerland, people queued up for hours to get a seat. Everyone who gathered for the announcement almost certainly knew the outcome of the experiment beforehand; it simply increased their appetite to be present. Knowing the end of a story need not spoil the enjoyment of it being told.

If, hypothetically, the CERN scientists had announced that they were giving their results and those results had not revealed the presence of the Higgs boson, the Audience would have felt let down; their expectations thwarted. The inverted pyramid Structure may seem like a narrative spoiler, in that it gives the game away at the very beginning of the communication, but it does effectively manage the Audience expectations. It is important to do this if you care about the Affect that you will have on your Audience. If, at the beginning of your report, you acknowledge that the results are not what you were expecting or hoping for, the Audience expectations from the Change and Affect of the communication are redirected.

The Middle

The beginning of a narrative draws the Audience in, sets the scene and introduces the puzzle. The middle of the narrative is the business of unravelling the Plot. In factual narratives it is the sequence of events which explains, sometimes in great detail, how the Obstacles are overcome. It is the part of the narrative which, because it requires most concentration, requires that the Audience is already interested and involved with the story. In the Eiffel tower film the detailed process of how the new lifts worked using electrical induction was presented once the viewers were engaged in the challenge: keeping the lifts open in all weathers. This was more than halfway into the narrative. If the film had begun with an explanation of electrical induction, many in the Audience would have switched channels. The principle is: delve into detail only when the Audience will be receptive to it.

Plots, as we have seen, can have more than a single Obstacle. Events and new discoveries bring to light new challenges which need to be solved. The question is how many of these need to become part of any communication. Perhaps this is best illustrated by the dilemma facing every biographer. Authors cannot possibly include every event in a person's life. The middle section is usually the longest section of a narrative; but it is not *that* long. Authors choose to illustrate a life with significant events which are important and relevant to the individual. Each life event has a Change and Affect associated with it which – separately and together – offer deep insights about their subject.

A narrative that unfolds chronologically may appear to have a simple linear Structure, but that does not equate to it being easy to construct. There is always more material than can, or should, be included. Decisions about what to leave out are always more taxing, in my experience, than what to include.

You take out the events and the details that do not contribute to the overall purpose of your communication. That it took you three years of fruitless effort is not relevant to an Audience unless, and this is entirely possible, you choose to make it a part of the narrative. Not the narrative about the results of the research, but a parallel narrative in a strand of its own.

Single Narratives in Series

Connected through an idea, theme or topic

The Romans were brilliant engineers and prolific inventors, so when we were making the television series *What the Romans Did for Us* about the science, technology and engineering methods introduced by the Romans to Britain, we were spoilt for choice over what to include in the series. Not surprisingly, we created a broad theme for each programme, in order to give us a filter for selecting stories. Our programme entitled *Arteries of the Empire* was fairly typical. It included six discrete stories. We covered not only road building and the surveying methods which the Romans introduced, but also one of the important reasons the Romans created the transport infrastructure in the first place: the gold mines in Wales.

We were able to create a series of stand-alone mini-narratives and we found a way to link one to the other using our theme. Between each mini-narrative, there is opportunity to pause and reflect on the story or to move swiftly on – perhaps with a change of gear or tone which suits the new narrative better. In each narrative there is an opportunity to enter a new World, be presented with a new puzzle, and therefore a different Obstacle to be overcome. The Change and Affect too will be different.

In the *What the Romans Did for Us* series we gave variety and pace to the programme by giving the Audience a range of experiences. Stories which demanded more concentration were placed next to those which made few intellectual demands. Our purpose was mainly entertainment, and our Structure reflected that. We took our Audience on an instructional and emotional journey, placing our stories in an order which amused, amazed, impressed and, on occasion, made the Audience think! The skill was in anticipating the Change and Affect of each story, and using the end of one story as a springboard to the next. The whole, that is all six mini-narratives, combined to provide some understanding of the technology which the Romans brought to Britain. Putting six mini-narratives together in an order that manages to achieve this is something which needs to be tried – or perhaps even discovered – through trial and error. We always designed the programme Structure on paper before we made each film, but once they were all shot, we usually rejigged their order. A better order of stories made for a more powerful overall Change and Affect, more *crackle and fizz* for the Audience.

Let us now look at an engineering example. This follows a similar Structure, but the connecting idea is a physical object: a bridge. The viaduct which crosses the Millau Valley in south-west France is an engineering wonder of the world. It is nearly two and a half kilometres long and it stands above the clouds most of the time, as it is the world's tallest bridge. It was designed by Norman Foster, a British architect, who, when the bridge was finished, said, 'I think the bridge is a fusion of the forces of nature, and extraordinary engineering expertise, and of a team working together'.

If, as an engineer working on the bridge, you were asked to give a presentation to young engineers about how the bridge was constructed, you could, no doubt, be guided by chronology and start with the first plans. Another approach was taken by producers of a British documentary who gave an account of the construction of the bridge, using five tales. Each one provided a back story of the first discovery and use of a critical piece of technology which made construction of such a vast bridge possible. So the story of how the Celts in ancient times split wood to construct their boats so that they kept their shape in all weathers, was the inspiration for the solution to the Obstacle of dealing with the heat expansion of the concrete deck on hot summer days. The GPS technology which the US Navy introduced to find their nuclear submarines when they rose to the sea surface became the solution to the Obstacle which the bridge builders had of maintaining millimetre precision over such vast distances.

Each of the five stories was told sequentially. The Big Hook for the one hour documentary was a puzzle put directly to the Audience. They were asked five questions including how a lost nuclear submarine and a crafty trick of ancient Celtic boat builders made the construction of the bridge possible.

Mystery, as we have seen, arouses Audiences' curiosity and gives them a reason to stay for the answer.

Sometimes a connecting thesis may not be obvious. It may need to be uncovered by looking at the potential material from different perspectives.

There was an example of this in the *Lure* chapter with the title of the lecture, *Confessions of a Leadership Coach*. The speaker told powerful, unrelated stories from his entire coaching experience, but his Structure linked them together. Each separate story was related to his confessions idea.

Connected through a purpose

At the beginning of my workshops I ask each person to introduce themselves. The guidance I give them is to describe three events which led to their presence in the room. I give them an end-point but not a starting point. They can choose where their story begins. It can start at their birth or with something which happened the previous week. They decide. My purpose in asking people to do this is to find out something about them in a way which encourages them to use a narrative format.

Most, but not all, provide three chronological events which are summaries of their high school education, their undergraduate degree and postgraduate research. Each story adds more information about the education and professional interests of the researcher. The Change that we understand is to their degree of specialisation. The Affect becomes memorable when researchers include a colourful anecdote about themselves which gives us an understanding of why they made particular choices. Those people whose purpose is to tell us more about themselves than their scientific interests alone tend to have more interesting narratives. Each story is structured to give us a different insight into their life, their personality or their science; whichever is most important to them.

Steve Jobs, the former CEO of Apple Computer and Pixar animation gave the commencement address to Stanford students in the summer of 2005.[4]

He told three stories in series with the purpose of revealing a philosophy about life which he wanted to share with the students. In each story Steve Jobs told the Audience something about the struggles of his early life, his disasters and successes with his career, and what it is like to face death. The speech was not a traditional curriculum vitae. Each narrative contributed to a vivid portrait of a bold, unconventional man with a deep desire to live life his own way and be in control of his own destiny. Each story's ending gave the impetus for the next story's beginning. His message to students was to follow their hearts and intuition. All his stories demonstrated ways in which he had done this. The whole was more than the sum of its parts.

The Complex Narrative

Narratives woven in parallel

The discovery of the stone sarcophagus beneath a car park in Bristol, introduced in the *Plot* chapter, was the beginning of a quest to find the answers to some intriguing questions: who was buried inside and what could we find out about him or her? In telling this story, as with any quest, we needed some leads, some clues which we could use to create a number of mini-narratives. We were not overwhelmed with choice. Quite the opposite, we were forced to use all of our creativity to find out as much as we could from what little we had: two skeletons, one of them incomplete, some teeth, a few rocks and a handful of hobnails. We needed to include everything that gave us some insight, however small, in our account. This gave us three mini-narratives. First, the skull gave us the opportunity to put a face to one of the skeletons. Second, the bones were investigated. We discovered one skeleton was male, in his mid-forties at his death and had ten times the normal concentration of lead in his bones. He had grown up in the Mendip hills and was probably a metal worker. The female was older than the male and placed in the coffin at a later date. The third mini-narrative was around the investigations which were triggered by the discovery of the hobnails. This gave us a story about the shoes which were worn by Romans towards the end of the Roman era in Britain.

On the face of it, these were not stories which, if told sequentially, were likely to grip three million television viewers. We had, however, some narrative skills up our sleeves which we used to keep our Audience with us. The Structure of our narrative was not a sequence of mini-narratives in series. Instead the mini-narratives were woven together in parallel to make a story stronger than any of its single components. In the middle section of the documentary, once the sarcophagus was rescued and opened and the skeletons were found, the mini-narratives unfolded in small sections. We moved from one mini-narrative to another and back again, creating gaps in knowledge and building suspense before revealing the answers.

Narratives as episodes

Sometimes there is so much material that it needs to be honed into a narrative which is tight and sharp, and not rambling. Biographers increasingly do not take a cradle-to-grave approach with their subjects, they use a different Structure: that used by Homer in his narrative on the Trojan war.

> '...[Homer] never attempts to make the whole war of Troy the subject of his poem, though that war had a beginning and an end. It would have been too vast a theme, and not easily embraced in a single view. If, again, he had kept it within moderate limits, it must have been over complicated by the variety of the incidents. As it is, he detaches a single portion, and admits as episodes many events from the general story of the war – such as the Catalogue of the ships and others – thus diversifying the poem.'
>
> Aristotle[5]

The biography is written through a focus on a particularly significant period in the person's life, with flashbacks. At various points in the story, biographers go back in time to tell of previous events in that person's life, which have a relevance to the 'significant' period.

Complex narratives are frequently used in fiction, films and novels, and in many documentaries. They are less widely used by specialists, except perhaps the most experienced, and those planning inaugural lectures. I have heard lecturers give an account of their research in parallel with the history of their subject. Others, particularly in medical research, interweave a strong patient story with the latest developments.

The CERN scientists who camped overnight outside the auditorium so that they could join the queue soon after sunrise did so, not merely because they wanted to be informed of the news of the discovery of the Higgs boson, but because they wanted to be part of history. They were highly involved in the quest for the Higgs boson, they had been on the journey, some of them for many years, and they had an overwhelming desire to be present to hear the public narrative, to experience a Change and Affect which they could tell their grandchildren about, and perhaps incorporate as an episode or a parallel story in their own lectures about physics and the Higgs boson.

Complex narratives which have a theme, for example a race or competition, allow for two or more narratives – one for each of the competitors – to be woven together. Exhibitions which show the work of two artists such as Picasso and Matisse, or the work of a neuroscientist alongside an artist, use narratives in parallel to create a richer Change and Affect experience for the Audience.

In narratives which run in parallel, there is usually a main Plot which follows the lead Character in search of an important goal, and one or more subplots sometimes involving the lead Character and sometimes someone different. Occasionally there are a succession of narrative episodes which reveal the history of a Character or a World through 'flashbacks' – narratives about the past.

Signposting

Creating a Structure which seamlessly entwines two or more narratives is not easy, as anyone who has sat through a movie that is confusing, a lecture that makes no sense or a book that is too easily put down, will know. The skill in designing a Structure with complex narratives comes in part, of course, from the strength of the individual stories. It relies, however, also on judgements about storytelling: when to introduce information, when to keep it back, what to include and what to leave out, when to switch to another narrative and when to switch back. It requires experimentation and practice. It also demands good signposting. The Audience need to know when the story is switching. Phrases like 'before I move on to show how we used this solution in our experiment, I need to introduce this piece of apparatus and tell you how we came upon it', orientate the Audience to go with the Structure you have designed.

And Finally

Another unifying device which holds a narrative together is the reintroduction or reappearance of earlier elements of the story. It can make a satisfying end to a communication when a recurring idea or detail makes another appearance. As an illustration I end this chapter, as I began it, with the thoughts of Michael Faraday whose advice to budding lecturers was extensive. He acknowledged the great difficulty with Structure. He wrote thus:

> 'I must confess that I have always found myself unable to arrange a subject as I go on as I perceive many others do. Thus, I could not begin a letter to you on the best methods of renovating our correspondence and, proceeding regularly with my subject, consider each part in order and finish, by a proper conclusion, my paper and matter together.
>
> I always find myself obliged, if my argument is of the least importance, to draw up a plan of it on paper and fill in the parts by recalling them to mind, either by association or otherwise. This done, I have a series of major and minor heads in order, and from these I work out my subject matter.
>
> Now this method... does well for the mere purpose of arrangement, yet introduces a dryness and stiffness into the style of the piece composed by it. For the parts come together like bricks, one flat on the other and though they may fit, yet they have the appearance of too much regularity. It is my wish if possible to become acquainted with a method by which I may write my exercise in a more natural and easy progression. I would, if possible, imitate a tree in its progression from roots to a trunk, to branches, twigs and leaves, where every alteration is made with so much care and effect that though the manner is constantly varied, the effect is precise and determined.'
>
> On the orderly arrangement of material[6]

The next and final chapter: *Make your Pitch CRACKLE* revisits the narrative ingredients and rules we have already met and uses a very particular Structure to engage and persuade the Audience.

Questions and Ideas to Help Stimulate Ideas for Creating a Structure for your Communication

1. The end: What is the final revelation of your communication? (This is where your narrative ends.)
    - If you are unsure where your narrative ends, have you defined your purpose?
    - What is the Change which the Audience will notice at the end of your communication? What Affect is that likely to create in the Audience?

2. The narrative form: If your communication is about your research, is there one single narrative thread which will create the Change and Affect you are planning, or are there opportunities to combine a number of mini-narratives to build on each other?
    - Were there events or Obstacles in your research project which warrant separate mini-narratives?
    - Can the mini-narratives be told in series, one following another?
    - Is there, perhaps, a significant piece of research which, at various places in the telling, you can pause to introduce flashbacks to earlier, pioneering work?
    - Is there a topic, a theme or a thesis which links the narratives, or could link them?
    - If the narratives are in parallel, is there a narrative device which will help to Structure the communication? For example: a race, a competition, a clash of cultures or a deadline?

3. Creating variety and pace: What are your ingredients? Use Post-it notes or postcards (or material which you have already gathered from previous exercises) to outline or draw the World, Big Hook, COFAD ingredients of the Plot, and the Change and Affect.
    - Consider the ingredients through filters of your choosing, for example tough to understand concepts or easier concepts. Heavy on mathematics or light on mathematics; parts which you find interesting or parts which you find less interesting.
    - If your research project was an Obstacle course, what was the pathway you took through it? Describe the journey as a narrative quest where the Audience are accompanying you. Where do you need a rest? Where do you need to prepare people for a big effort? Where are the dangers of losing people? What is the reward?

4. The beginning: What are your options for where you begin your communication?
    - Do you want to communicate the main purpose of your communication at its beginning? Is there a good reason for this?
    - Is there an intriguing observation, object or anecdote from the body of your communication, or from the end of your communication, which you can use as a Lure?
    - Revisit the *Lure* chapter for ideas.

9

Make Your Pitch *CRACKLE*: how to persuade, to win support

> 'Persuasion must in every case be effected either:
>
> 1. By working on the emotions of the judges themselves
> 2. By giving the right impression of the speaker's character
> 3. By promoting the truth of the statements made.'
>
> Aristotle[1]

There is a story in *Tom Sawyer*, the novel by Mark Twain[2], where Tom is ordered by his aunt to whitewash the garden fence. It is a punishment for some misdemeanour. Tom is discovered by his friends carrying out this 'melancholy' activity but is determined not to be humiliated. He reframes the activity from one of punishment to one of pleasure, and manages to convince his friends that white-washing the fence is a rare and thoroughly enjoyable activity. His friends are so persuaded that they not only line up to take a turn with the brush, but they offer him gifts for the privilege.

I loved that passage when I first read it as a child and I still do. A simple but masterful example of a Character, Tom, using his intelligence to manipulate others, transforms a group of potentially derisory friends into admiring ones and turns an unpleasant experience into a good one. Regardless of the morality of the actions, it is an excellent example of effective persuasion. You do not have to be right or good or honest to be a powerful persuader (although obviously it is better for everyone if you are), you need only to be convincing.

I learned to persuade at a professional level, that is to Pitch, at the beginning of my career when I worked on the BBC science magazine television programme *Tomorrow's World*, introduced in the *Structure* chapter. Every other week we were expected to produce a short story (an item) for this half-hour show fitting to the programme brief, which was known to anyone who watched the programme. The story needed to have at its core some science, medicine or technology, preferably something which could be demonstrated in a live studio. The item could be from any area of home, work or industrial life. An individual programme would have as many as seven items on a variety of topics. A single show might include stories about a bridge, a helicopter, a new gadget for the kitchen, a bullet-proof vest and a treatment for diabetes. One of the most important attributes of the item was that it had to be about something novel; a piece of science or engineering which would be new to our Audiences. That is, we wanted to produce stories which only a handful of the eight-million strong Audience had heard of or read about before.

We found these stories in a number of different ways: scouring the trade magazines, going to exhibitions, visiting university research departments and reading scientific journals. Finding the stories, however, was the easy part. Persuading colleagues that the story was worthwhile and fitted all the criteria for the programme brief, was more difficult.

Each Friday following the Thursday-night live programme we reviewed the show and afterwards we pitched the new stories we had found to the gathered team. It could be a bruising experience for the unprepared newcomer, as I remember well.

If the story was not pitched well, however strong one personally thought it was, the assembled group would shake their heads or show a 'thumbs down' sign to communicate their view that it was not sufficiently good to merit inclusion in the show. Equally obvious to the newcomer was that when stories were pitched well, even if they later turned out to be a little thin on content, they were quite often given the thumbs up. They also tended to be approved when they were pitched by one of the senior members of the team.

I learned very quickly, as most of the survivors of this intense training did, that to Pitch to this group of people, all competitors for the limited space in each programme required skill – one that was acquired with practice. I also learned that sometimes it was best not to Pitch some stories at the Friday meeting at all, but to wait until the editor was short of an item for that week's show. Pitching to an individual, whose needs are urgent and known is much easier than to a group whose needs are complex and less straightforward.

Pitching is an essential skill for anyone who has to compete for resources. In science there is fierce competition for grants, for jobs, for space at conferences, poster presentations and in the 'high impact' journals. There is also competition for venture capitalists' backing.

In the UK, publicly funded academics are now required (as part of the Research Excellence Framework (REF)) to make a persuasive case to show evidence that their work has a social, economic or cultural effect on society. In some universities these 'impact statements' – as they are called – are written, not by the scientists, but by people in the marketing department. This is a clear acknowledgement by the research community that professional pitching techniques are needed to ensure their research is communicated in the most effective way.

This chapter is about making your thoughts and ideas clear in order to persuade people that your work is of value. In other words, it is about how to Pitch ideas and points of view, and how to sell yourself.

None of the narrative terms that we shall meet in this chapter are new. They have been discussed earlier in the book. What is new is the way that they are put together in a Pitch.

The intention is to expand on Aristotle's advice on persuasion with a framework which encapsulates the essential ingredients of a Pitch. The framework has the acronym *CRACKLE*:

C **Challenge** – What is the Challenge which you, or your idea will address?
What is your goal?
What is your opportunity?
What is the problem you want to solve?

R **Reasons to Care** – Why is your Challenge important, not to you, but to your potential sponsors?
Why do they care – or why might they care? Why now?
How would solving the problem help them?
Why might your opportunity be of interest to them?

A **Actions** – What do you propose to do in order to complete the Challenge?
Actions are the guts of your idea. They are your implementation plans.
They are what you intend to do to be successful in achieving your goal, meeting your opportunity or solving your problem.

C **Character** – Why are you (or your team) best placed to deliver success?
What do you bring to the Challenge or project which is unique?
What qualities do you have which show that you are the best person for this job?

K **Key Benefits** – What are they? How do they deliver value?
How will your idea be of benefit to the sponsor?
Can you quantify the benefits?
Can you show that overall benefits are of greater value than any losses which might be incurred?

L **Lure** – To get attention.
Do you have any surprising facts or figures?
Is there a question you can pose to raise the interest of your sponsor?

E **Ending** – A powerful close which persuades the Audience to say YES!
A question which defines what you are asking of the sponsor.

My intention is to demonstrate that a Pitch made with *CRACKLE* has clarity and persuasive power.

Following an explanation of each of the *CRACKLE* ingredients, I have included a questions section related to the individual ingredients. Along with the standard Pitch questions relevant to all persuasive proposals there are questions to help you prepare a

CRACKLE Pitch for grants and funds, an interview or job application, an impact statement, persuading your boss to let you attend a conference halfway across the world or getting permission for a colleague to join you in a project. Skip over the questions which are not relevant to your situation.

We shall also consider how to improve on a Pitch. At the end of the chapter there are some examples of Pitches, including what is known as an 'elevator Pitch'. This derives from the idea of having to make a succinct Pitch to someone influential during a one-minute ride in an elevator.

Your Audience – *The gatekeepers*

When people start out to plan a Pitch, most have a belief in the strength of their idea and assume that, once it is outlined, others will see its value as they do. Sadly it usually requires more than this. Persuading individuals to become cheerleaders for someone else's idea requires skill.

The starting point for thinking about a Pitch is the same for any communication: the World of the Audience. In this case the Audience is a sponsor or, as we shall refer to them in this chapter, a gatekeeper.

Gatekeepers are the people who control access to scarce resources through their decisions. They can be an individual – a venture capitalist with money to invest, a professor with limited jobs to offer or a brilliant PhD student who wants to do worthwhile research. Gatekeepers are often members of a committee acting in the interests of a funding body, a charity, a commercial company or taxpayers. In television, they are the commissioners and editors of programmes working on behalf of the Audience.

Gatekeepers are critical players in your Pitch communication. Effective persuaders get to know their gatekeepers. They want to know all about them – their likes and dislikes, their strengths and weaknesses, their career choices and past decisions. In extreme cases they find former and present colleagues of the gatekeeper and listen to the stories which people tell about them. The more they can understand about what motivates the gatekeeper, the better able they are to tailor a Pitch that comes across to him or her as irresistible.

Gatekeepers are the primary Audience for your Pitch. So here are some questions to help you begin to try to understand the World as they see it.

The individual gatekeeper

Someone who makes decisions alone. Those decisions could be:

Who to work for?
Who to promote?
Who to collaborate with?
Which project to invest in?

1. What is the best approach to take with this gatekeeper?
   - What kind of person is the gatekeeper?
   - Where does he lie on the risk spectrum? Does he follow hunches, take chances or is he risk-averse?
   - Is he more introverted or extroverted; confident or less confident?
   - What are his preferences for assessing new ideas: through reading, listening or discussing with others or perhaps taking part in open debates?
   - What is his position in the hierarchy at work? How does he interact with others, such as his peers and those in more senior and junior positions? What does this tell you about his needs?
   - Also consider when is the best time to approach the gatekeeper?

2. In a typical day how does the gatekeeper allocate his time?
   - What does this tell you about what he thinks is important?
   - Whose opinions does he value?

3. What does the gatekeeper really enjoy about his work?

    - Is the research focus of paramount importance? If so, what is important about it from the gatekeeper's point of view?
    - Is the opportunity to work with like-minded souls critical?
    - Is the chance to learn from others or perhaps to develop others important?

4. What does the gatekeeper *not* enjoy about his work?

    This question can be as useful as question three in helping to tease out the strengths and weaknesses of the gatekeeper.

5. What stories have you heard about the gatekeeper?

    - What do they tell you that is new? Does it fit with what you already know? Are they reliable? Can you check them out?

If you want to persuade a boss to give you funds for a conference or piece of equipment, or if you want to persuade a coworker to join you, here are a couple of additional questions:

6. Can you envisage the gatekeeper making a decision in your favour?

    - Are there any precedents that you can learn from?
    - What, if anything, would need to change for your idea to fit better with his Challenges and his needs?
    - How could you help that happen?
    - Who else would be affected by this change?
    - How can you gain their support or agreement?

7. What will change for the gatekeeper if he becomes a champion for you, your project or idea? List the answers to the gatekeeper's question, 'What is in it for me?'

    - What will he gain? Are these gains important things that he wanted or needed?

- How might your proposition make him feel good about his decision?
- Will it make him look good in the eyes of his colleagues?
- What could he lose?
- How could you lessen the loss or increase the gain?

The gatekeeper representing others

The gatekeepers are members of a panel or committee. They are collectively responsible for their decisions which could be:

Who to award grants to.
Who to publish.
Who to promote.
How to allocate the time and the space at a conference.
How to assess the quality of REF submissions and impact statements.

Decisions are broadly made against a set of criteria, although clearly personal preferences of the individual gatekeepers can be a contributing factor.

1. What are the decision-making criteria used by the panel?

   - How are they different from previous years (if a conference)?
   - In what ways are they different from other organisations offering similar opportunities? Why do you think they have this difference? What does it tell you about their priorities?

2. What kinds of projects has this group funded, published or made successful in the REF process in the past few years?

   - What advice can others, especially those who have been successful or unsuccessful with this committee, give you about the application process?
   - Have you looked at the funders' website, recent copies of the publication you are applying to, or the REF website[3] for

examples of case studies from the impact pilot exercise? What can you tell about the gatekeepers' decisions from these?

3. What else can you learn from the website, the application form and other published material about the culture and values of the organisation?
   - Does the organisation produce a regular magazine or newspaper? What kind of research, people and institutions are featured? What does this tell you about the organisation?
   - Have there been any reports or news items in the National or International press about the organisation? How might these reports affect the gatekeepers?
4. What are the backgrounds and research interests of the members of the committee?
   - What does this tell you about the ambition and priorities of the organisation?

The *CRACKLE* Components

Challenge

The Challenge might be:

A problem to solve
An opportunity to exploit
A goal to achieve.

It is the Big Hook puzzle in a narrative and the brief which outlines a project.

> The Challenge is often a gap of some kind: a gap in understanding, a gap in opportunity, a gap in resource, a gap in need or desire.

The Challenge is the problem or opportunity which you are addressing. It is that simple. It needs to be very precisely defined because, as we discovered in the *Big Hook* chapter (page 102), the Challenge or problem you present must be the one which you will solve.

When you are looking for support for tackling your Challenge, you need to find ways to show that your goal, your problem or your opportunity is relevant to your potential gatekeeper–sponsor. This might be relatively easy, especially if the gatekeeper you are applying to has set the Challenge and it is a simple one. An advertised job, for example, might have a clear job description. This is the Challenge as the employer has defined it. It is your starting point for considering the Challenge for your Pitch for the job.

Competitions are set with Challenges which are clear and unambiguous. *The Great Egg Race,* a television programme I worked on in the 1980s, set its first Challenge to engineers across the UK. The goal was to create a vehicle to transport an egg over the furthest distance. (Each vehicle had a single rubber band, issued by the sponsors, as the energy source.)

More often the Challenges set by gatekeepers are more complex. Science funders frequently call for ideas from multidisciplinary teams to tackle big global Challenges – for example, how to solve the problem of water pollution, how to feed the world's population, how to cure Alzheimer's disease, and so on.

When you are pitching projects to get access to these funds, you will need to do more than acknowledge or repeat the gatekeeper's Challenge. The task is to redefine the Challenge such that it addresses a more precise need, which consequently indicates how the main Challenge might be met.

Let me illustrate this by revisiting the example we met in the *Audience* chapter, where advertising planners were set the broad Challenge of making a film to persuade busy fathers to read to their children. The advertising planners worked on the Challenge they were set in order to develop a more focused, realisable Challenge: to show fathers *how* they could find time to read to their children. This more precisely defined Challenge both addressed the needs of the fathers and hinted at the manner in which the Challenge would be addressed.

If you are seeking funds from a sponsor whose main aim, for example, is to find a cure for malaria, the Challenge you express in your Pitch needs to encapsulate the problem – or opportunity – which you have identified as the route to getting nearer to a solution. For example, your Challenge might be to interfere with the *Plasmodium* parasite's life-cycle, or to investigate the sex life of the mosquito. It might also be to look at another related parasite, such as *Toxoplasma gondii*, which can act as a model for *Plasmodium*. Your Challenge is likely to have a narrower focus than the gatekeepers' Challenge, but it must still resonate with the bigger Challenge. The gatekeepers should recognise it as a Challenge they share; a Challenge that, if solved, will satisfy their needs.

There are many Challenges and opportunities in this world which go unrecognised or unstated until someone has an idea. With a flash of inspiration, someone sees a way through a problem or stumbles upon an opportunity to make a difference. This is the domain of innovators and entrepreneurs – people who regularly Pitch for money and support. Innovators – scientists and engineers are often innovators – have ideas to solve problems which may never have been articulated. They approach gatekeepers with more than a clear definition of their intention. Not only do they define their Challenge precisely, but they find reasons why the gatekeepers will care about it.

Here are some questions to help you frame your Challenge:

The Challenge questions

1. What is the Challenge?
   - Does your Challenge define the problem or opportunity with sufficient clarity? What is special about the Challenge as defined in your proposal? Why is it pertinent now?
2. Is it a gap that you are attempting to fill?
   - What is the gap in understanding, in knowledge, in capability, in technology, in ability, in the market or in desired behaviour? Is there any other gap you have identified?

- What opportunity or insight have you uncovered?
- What is currently missing – that is not available, but could be made available – from the World of the gatekeeper?

3. What are your assumptions about the Challenge?

   - What are the assumptions that people (gatekeepers, or the general community or public) make about the Challenge? Do these assumptions need challenging?
   - Ask yourself what completing the Challenge would allow you to do? Does this reveal another, potentially more important, Challenge that should be mentioned?

4. What is the evidence that this is an important Challenge for your gatekeeper?

   - Where does this evidence come from?
   - How reliable is it?
   - Have you tried looking for evidence from a number of different perspectives? For example, from financial, health, scientific, employment, global or any other relevant associations?
   - Is there evidence for a bigger Challenge – or a smaller, more manageable one – which your idea relates to?

If the Challenge is a job opportunity, can you (from the job advertisement) identify the critical goals, opportunities or problems which form the primary Challenge of the gatekeeper?

Reasons to Care

Let us deal first with the Reasons why *you* might care about your Challenge and want to pursue it. You will have a mixture of intellectual and emotional reasons. You may believe – and indeed have evidence to support your belief – that your idea makes logical sense:

It could deliver a solution to a big problem
It could save lives, or time, or money

It could offer something quite new and quantifiably useful to people, to industry, science or technology.

In the *Audience* chapter we introduced Arthur Maslow's hierarchy of needs, the human motivations which apply to all of us. These will be among your other Reasons to Care about your Challenge:

An interesting or exciting project which challenges your intellect
Gives you job security
Gives your career a boost
Makes your job easier
Gives you wealth
Enriches your life
Gives you status or provides esteem

None of these purely personal Reasons to Care are relevant to a Pitch. Not even if they are your prime motivation.

Think back to Tom Sawyer whose Challenge was to persuade his friends to see him as 'cool' and not a 'loser'. His Reasons to Care were completely self-interested, namely that he did not want to lose face. That, however, was not the Challenge he presented to his friends. He reframed his Challenge to appeal to their particular needs: the opportunity to whitewash a picket fence as a highly sought after experience. This revised Challenge was consistent with his, and it gave him a way of thinking what he needed to do.

> 'He had discovered a great law of human action, without knowing it – namely, that in order to make a man or a boy covet a thing, it is only necessary to make the thing difficult to attain.'
>
> Mark Twain[2]

The same principle of thinking about the needs of Tom's friends applies to articulating the Reasons to Care in a Pitch to gatekeepers. When the Challenge is shared, the Reasons to Care are likely to coincide. When your Challenge is personal, you will need to work hard to find reasons why other people should care about it and why other people will believe it to be as important as you do.

Tom Sawyer presented his Reasons to Care like many a sales representative: a rare opportunity with time-limited availability.

In a Pitch or any persuasive narrative, the Reasons to Care which you present to the gatekeepers must reveal, unless they are completely obvious or intuitive, why they are important to them and their interests. Venture capitalists will have Reasons to Care if you can demonstrate an exciting gap in the market and an opportunity for them to make a return on their investment. Coworkers will have Reasons to Care about joining a research group that offers them the opportunities for publishing joint papers, for potentially getting a promotion, for the chance to move to a new city or for other, personal, reasons. A boss' Reasons to Care to send someone to attend a conference might be to have his team represented, to present the latest results or to give a personal reward to a particular researcher.

Reasons to Care about the Challenges in research are also a necessary part of persuading people that the Challenge is worthwhile. Funding bodies and REF committees may not need a financial return on the grants they allocate, or the work they approve, but they are looking to support research Challenges which are important. They want their money to deliver something of scientific significance which might have an impact on wider society. Funding bodies look to experts' opinions to confirm that there are sufficient Reasons to Care that the Challenge and the specific gap in knowledge, outlined in a research project, is important and meets the bigger Challenge set out in the goals of the grant giver.

Many inexperienced pitchers focus all their efforts on their proposed Actions. They do not spend the time they should thinking about the Challenge and the reasons why others will care about it. Actions – as with any narrative Plot – only become intriguing to an Audience once they care about the central Big Hook Challenge.

Here are some questions to help you bolster your Pitch or impact statement with supporting facts and arguments to show that the Challenge you have defined is an important one.

Reasons to Care questions

(The Challenge might be to gain funds, to go to a conference thousands of miles away, to get a job, to get a coworker to join forces with you or to support your idea.)

1. Why do you care about the Challenge?
    - Is this the same reason that others may care?
    - What other reasons are there to care?

2. Put yourself in the position of the individual gatekeeper and ask yourself:
    - Why do I care about this Challenge – problem, goal or opportunity?
    - Why could it be important to me or for me?
    - Why now?
    - What is in it for me if I give my support? What is in it for people I am responsible for?
    - What else would it take to make me want to give my support?

3. Put yourself in the position of the gatekeeper on a committee and ask yourself:
    - Is this Challenge mainstream enough? Will it help to solve an important problem?
    - Is this something which others on the panel will support? What else would it take for others to agree to give their support?

4. What is the evidence?
    - Is there evidence to support why the gatekeeper might have Reasons to Care?
    - Where could you find the supporting evidence?
    - Do his Reasons to Care offer you an opportunity to reframe the central Challenge?

3. If you could succeed in the Challenge, what would Change?
    - Does this list reveal any additional Reasons to Care?

Actions

Actions are the guts of your idea. They are your project plan. They provide gatekeepers with the details of what you intend to do to succeed – and how you mean to do it. In narrative terms, Actions are your Plot.

> Your aim in a Pitch is to explain clearly your intended Actions and show how they will succeed in meeting the Challenge.

If you have defined your Challenge well and given your gatekeepers sufficient Reasons to Care, they are likely to feel some ownership of it; their minds will *crackle and fizz* with possibilities as they assess your Action plan.

Actions, your direct response to a Challenge, should come across to the gatekeepers as relevant, interesting and engaging. Tom Sawyer's Actions were mostly to act and feign a passion for a task he was not enjoying. Actions for a Pitch for a new job could be to outline your plans for the first 100 days in the role. Setting up women's groups could be your headline Action in a Challenge to reduce maternal mortality in Nepal. Making a short film to illustrate the impact of your research on patients' lives might be an Action to publicise your work or to persuade funders to give further support. Running a pilot project might be an Action to determine the feasibility of the underlying idea.

The Actions in an impact statement are what you *did* to meet the Challenge, and how you did it. In both a Pitch and a REF impact statement, the gatekeepers not only want to have confidence that your plans are, or were, creative and are likely to work, they also want to know whether you, the pitcher, will be a Character who can deliver a successful outcome.

Here are some questions to help you think through the Actions which will achieve your ends.

Actions questions for Challenges involving funds or grants

1. What are your plans for success?

   - Have you done a project plan laying out your planned Actions along a timeline?
   - Have you worked out what your project plan will cost?
   - Do you have a contingency plan if something unexpected occurs?

2. What evidence do you have that your planned Actions will meet the Challenge?

   - Have you talked through your plan with someone who has done something comparable?
   - What could go wrong? How will you make sure it doesn't?

3. What kind of Obstacles or problems are in the way of success?
   - Are they technical, practical or financial?
   - Do they involve resources or people?
   - If you identify individual people who appear to be unsupportive, then use it as a Challenge. Use the questions in *The individual gatekeeper* and the *Reasons to Care* sections to help you find ways to influence the people you need support from.

4. Has the gatekeeper tried before to overcome these Obstacles?

   - What happened? What can you learn from his experience?

5. Has anyone else tried to overcome these Obstacles?

   - What can you learn from previous attempts to overcome the Obstacles?
   - Has anyone come up with a solution to a similar problem, perhaps in a different context? If yes, then is more research necessary?

6. Have you developed something novel (either technically or perhaps a novel approach) which will meet the Challenge?

   - What is it?
   - How does it work?
   - Why has no one done it before?
   - How did you develop it? Has it been tested thoroughly? Are you confident it will work?
   - Have you experienced success with a similar Challenge? What did you learn which you could apply to this situation?

7. What are the other important Obstacles?

   - Have you thought through all the important Obstacles that are likely to come up as you plan your journey from proposition to implementation?

Actions questions for refining impact statements

8. What kind of Obstacles or problems were in the way of success?

   - Were they technical, practical or others?
   - How did you overcome them?

9. Had anyone tried before to overcome these Obstacles?

   - What happened? What did you learn from their experience?

10. Did you develop something novel (either technically or perhaps a novel approach) to meet the Challenge?

    - What was it?
    - How did you develop it? How did you test it?

Actions questions for interview preparation

11. What are your plans for the role?

- How do they meet each of the Challenges outlined in the job description?
- Are there other Actions you have thought of which would be of benefit to the gatekeeper?

12. Have you taken advice from members of the gatekeeper's team or coworkers?

Character

When anyone said they had a brilliant idea, a colleague of mine used to say – on a regular basis – 'Any fool can have an idea; it is what you do with it that matters'. This is not an encouraging response to someone excited by their proposition, but I have come to realise over my career that there is absolute truth in this statement. Ideas alone have no value. They require people to implement them.

New ideas need champions who are prepared to struggle to make them a reality. Champions are the scientists, engineers, researchers, investigators, innovators and the people who bring ideas into the world by rising to meet Challenges. In narrative terms, they are the Characters who drive the Actions. Part of the role of the Character is taking Action to win support and resources, so it is often they who will perform the Pitch. (Not always. In marketing companies there are professional pitchers who Pitch the idea, knowing that others – if support is won – will perform the actual task.)

If you think of yourself as a Character in the Pitch, then you are the hero. When gatekeepers are deciding whether to join or support a Character, they want to feel confident that the Character is someone deserving of support, someone special who will follow through with their Actions and deliver what is proposed. The Character needs to demonstrate a number of qualities:

- *Knowledge and understanding* of what they are proposing.
- *Expertise and experience in the area of activity* or, at the very least, access to it. Characters who can show a track record of

delivering similar kinds of project are simply less of a risk to a gatekeeper than those with less experience. Gatekeepers who like to play safe place a higher value on experience than on a radical idea.
- *Passion*. Gatekeepers look for passion because it shows commitment, an attribute which they know will be required to keep the Character's spirits up when the going gets tough, as surely it will in most projects.
- *Integrity*. Gatekeepers, like everyone else, want to be able to trust the people they work with.

Gatekeepers also want to know that they are working with the best team for the Challenge. If you can bring something unique to your project, such as access to a rare and necessary piece of equipment, or colleagues with specialised skills in a particular field of study, then you can exploit these in a Pitch.

If these unique attributes are missing, it can be more difficult to win support. However, many gatekeepers are willing to support a Character who shows an ability to work in a team or is amenable to developing ideas with contributions from the gatekeeper or others. Gatekeepers are primarily concerned with getting the best deal for their investment or for themselves. They look to the Character of the pitching team as just one of the Key Benefits of the idea. There will need to be others too.

In preparing a Pitch, ask yourself what it is you are bringing to the project that is unique and superior to others who will be rivals for funds. Here are some questions to help you prepare.

Character questions

1. Why you?
   - What is most important to you about the idea? Why? What does it say about you?
   - What have you done before that is relevant to this project?

- Have you failed in the past in a similar situation? Why was that? What did you learn? Can you do something different this time to make sure that you don't make the same mistakes?

2. How will your personal contribution bring rewards to the gatekeeper for his support? (For an impact statement, how did your personal contribution make a difference to science, culture or society?)
   - Are there resources or people you can draw on for support?

3. Can you recall an experience where you had to make a difficult choice?
   - If the experience demonstrates a strength, such as your courage, perseverance or creativity, then consider if it would be appropriate to mention as support for your case with a venture capitalist, a potential coworker or at a job interview.

Key Benefits

This is how you address Aristotle's dictum to 'promote the truth of the statements made', particularly as they relate to the Change and Affect of your proposition.

Gatekeepers, unless they are extreme risk takers who like to work on hunches – and very few of those are long-term survivors – like to be as certain as they can be that their investments will be worthwhile.

> Key Benefits provide the evidence that your Actions will deliver overall value to the gatekeeper.

They are a quantifiable response to the Reasons to Care.

The purpose of outlining the Key Benefits, showing precisely how your idea will deliver value, is to show the gatekeeper that you have addressed the risks of your Actions. That is, you have considered all those Obstacles which could stop you being successful.

It is very easy to make optimistic statements about how an idea will appeal to its intended Audience, such as 'people care about the environment, so they will share cars to work if they can do so for free'. A Pitch needs to be more rigorous. It should give estimations, based on current data, of the potential demand for the idea, of its costs and of its impact. The Key Benefits show where the idea brings value.

The value of something is its 'measured want':

How many people will benefit from the idea?
In which way will they benefit?
Can the benefit be expressed in savings to cost, to time, to efficiency or even in lives saved or made better?
What will people pay for a product? How many will buy it?
Are the projected costs of the project less than the potential returns?

When you are preparing the Pitch you may not have gathered all the evidence of the Key Benefits. Indeed, some of the benefits may not be known. The point of *CRACKLE* is to help you focus on where your case is weak so you can take steps to find the evidence which will give your idea more credence. Or, it might be to accept that in certain cases – pure research projects for example – there may be no evidence that the Challenge being pursued will have any Key Benefits beyond the pursuit of knowledge. In these cases the Reasons to Care about the Challenge might be limited.

Physicists cared deeply about the search for the Higgs boson because it was an important Challenge which represented the one remaining gap in their Standard Model of nuclear and particle physics. This, however, was enough to get it funded.

Key Benefits questions

1. Who benefits and how?

    - Who benefits from the success of the Challenge? What evidence is there for this? How do they benefit?

- How does the gatekeeper benefit? What is your evidence for this?
- Why will the gatekeeper choose to back you rather than a competing idea or person?
- What evidence do you have that your idea has more value than others tackling the same Challenge?
- Are there any special or unique benefits which you or your idea will deliver?

2. Can you quantify the Key Benefits?

   - How can you quantify the value of the Key Benefits? (Remember, value is a *measured* want.)
   - Can the benefit be expressed in costs, in effectiveness, in time, in efficiency or in lives saved?
   - Where would you find the supporting evidence to present the Key Benefits this way?

3. What is your evidence for commercial success?

   - What will people pay for a product? How many will buy it? Where is your evidence for this?
   - Who will sell it? How will they hear about it? How will it be distributed?
   - Are the costs of the project less than the return?

Lure

The Lure does the same job that it does in any narrative: attracts the attention and interest of the Audience, in this case the gatekeepers. It is the only ingredient which is always at the start of the Pitch or, if it is a poster, the big headline or central picture.

Suggestions for creating a Lure

Choose a Lure from the nuggets of facts or stories you have accumulated for *CRACKLE*. Choose one which is surprising, relevant, intriguing and guaranteed to catch the attention and the interest of your intended Audience.

The Lure can be a compelling fact, directly related to the nature, scale or severity of the Challenge, for example, *200,000 people in the UK are affected annually by heart disease.* It can also be a question: *Do you know how many people in the UK are affected by heart disease each year?*

The Lure can be a brief touching story, a Reason to Care about the Challenge: *Anna is seven. Without help she will die from a completely preventable disease...*

The Lure can be a prop or a model of something which your Actions will produce. Bringing a prototype to a Pitch can help make the project seem 'real': *My team has designed this: a clear synthetic short tube with three valves which mimics the action of one of the four valves in the heart* (Figure 9.1)[4].

The Lure can be an intriguing piece of information specifically about you as the Character: *My research group has made a synthetic heart valve out of blocks of special polymers which self assemble into the desired pattern as a result of our proprietary processing technology.*

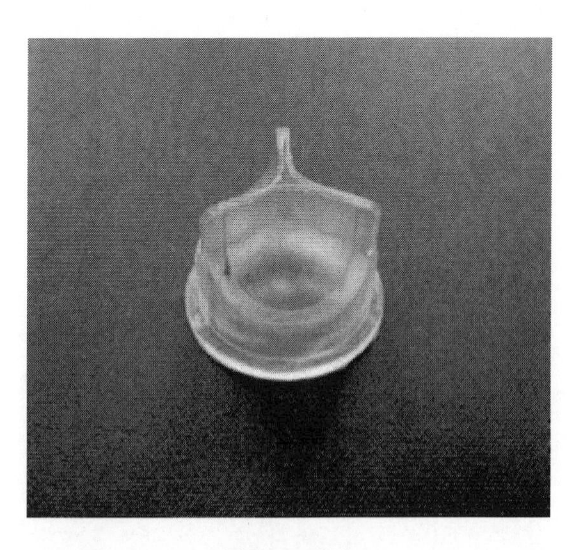

Figure 9.1

Ending

The Ending is where you conclude your Pitch, making it clear to the gatekeepers what is expected of them.

> The Ending should be a powerful close or a 'rallying cry'.

A Pitch (unlike an impact statement, for example) needs a strong Ending where you make it clear what you are expecting of the gatekeeper or Audience. Alternatively, and this is more usual for the short elevator Pitches, your close is a request for a meeting to discuss the idea further or to send the full proposal for consideration.

Endings, some examples

A direct question, a request or a clear statement of what you need.

- 'I have written this four-page value proposition, including some more details. Can I book an appointment for us to discuss it next week?'
- 'Are you free now to talk it over further?'
- '... *that* is what I propose to do if elected. Please vote for me!'
- 'I am wanting to start work next month. Would that suit you?'
- 'We believe we have a great idea, a great team and a clear vision. The only thing that's missing now are mentors and the best location from which to launch so we'd love the opportunity to come out and meet you!'

Putting it All Together

CRACKLE is a narrative framework which has the kit of parts required to construct a persuasive proposition which has, above all, clarity. Each Pitch must be constructed with particular gatekeepers in mind, so the same idea may use a different definition of the Challenge and provide different Reasons to Care and outline different Key Benefits. For example, a Pitch to a cancer charity to fund a project on cell division is likely to be a variation of your Pitch to a general science funding body. It will be significantly different from a Pitch you might make to the organisers of a science festival to bid for a slot to talk about your work.

As with other narratives, there is no set order for revealing the separate ingredients except for the Lure and the Ending.

The Improvement of the Pitch

The main advantages of a well-planned Pitch are obvious. If you know what you want to say, and have rehearsed it so that you are confident it is clear, concise and makes your argument as strongly as the evidence allows, you will avoid the hazard of making *ad hoc* choices about content, which might do little to help your cause. Instead you will be able to concentrate on the delivery of the Pitch.

> 'The first draft of anything is always shit.'
>
> Attributed to Ernest Hemingway

The best way to improve your Pitch is to subject it to rigorous scrutiny. Ask critical questions of each of the *CRACKLE* ingredients. By far the most efficient way to do this is with a group of trusted critical colleagues. They can be people who have a vested interest in your idea as well as people from different parts of your institution or business who are prepared to give you honest, supportive feedback. In the workshops I have run at universities, it is usual that people from different disciplines help improve each other's pitches by giving supportive, constructive feedback and asking useful questions. Some have found this so helpful that they continue to meet when someone has an important Pitch to make. Quite simply, it is a profoundly efficient way of improving each draft and I cannot recommend it more highly.

By inviting people who are not necessarily experts in your subject, you are required to make the verbal Pitch comprehensible and interesting to non-specialists. It is something which many funding committees now demand from researchers by asking for a lay summary.

And Finally, Some Examples

An example of an advertising Pitch

If you travel on public transport in any large city you will have noticed some small poster advertisements, mainly for charities, which ask travellers to text a number to make a donation – usually a few dollars or pounds. These advertisements are making a Pitch to commuters. They contain the *CRACKLE* ingredients as a mixture of pictures and text. In the example below, notice how the big Challenge of the charity – to raise funds to help refugees – has been worked into a smaller Challenge for this particular advertisement. This gives commuters a sense that the Challenge is manageable. They can contribute something small, yet make a real difference to an important cause. The *CRACKLE* ingredients have been honed to appeal to commuters.

C The **Challenge** is to protect a child refugee from bitterly cold nights.

R The **Reasons to Care** are that the child is a powerless victim and keeping her warm at night might save her life.

A The **Actions** being proposed is commuters can donate money to provide a blanket.

C The **Character's** strength is that the charity has links with the refugee camps and can deliver the blanket.

K The **Key Benefits** are that the blanket will keep the child warm, preventing her dying from cold.

L The **Lure** is the large endearing photograph of the child looking directly into the camera.

E The **Ending** is asking commuters to send a one word text message by mobile phone to donate just £3 to buy a blanket.

An example of an elevator Pitch

Dr Geoff Moggridge (GM) is a Cambridge scientist who works with smart materials. His team has designed a new kind of synthetic heart valve which one day might replace faulty human valves. This is his first draft of an elevator Pitch that he made to a group of fellow scientists, acting as a funding body. He has kindly allowed me to use his Pitch as an example. It includes all of the *CRACKLE* components.

> Every year, 200,000 people with heart disease need to have a heart valve replacement. At the moment if you have a heart valve which needs replacing there are two options. Firstly, you can have a mechanical device fitted. These work well, but the downside is that you will need to take warfarin, a blood thinning agent, for the rest of your life. The second option is you can have a biomimetic valve fitted, which is crafted from the pericardium tissue of a pig or a cow.
>
> These work well for older people but are not a realistic option for younger active patients as they do not last more than ten years. This is because the tissue is dead and in a different location to that which it grew.
>
> I have designed this (*GM holds up his prop*). This is a synthetic heart valve which my research group has made as a block copolymer structure, that is, blocks of different polymerised monomers oriented so that the fibres mimic the strong, flexible, fibrous material of the natural heart valve. The fibres are each less than one-thousandth the breadth of a human hair and self assemble into the desired pattern as a result of our proprietary processing technology. The material expands like a balloon but has the strength of a horse tendon. I believe that this new nano-porous fibrous material could be used to produce alternative better heart valves.
>
> I would like to ask you for £100,000 to buy an accelerated heart machine which will test the synthetic heart valve over a lifetime in just one year.
>
> Can I come and talk to you further about this?

C The **Challenge** is to improve on the mechanical replacement heart valves which are given to younger people.

R The **Reasons to Care** is that the younger active patients with a mechanical heart valve need to take a potentially toxic substance: warfarin, for the rest of their lives.

A The **Actions** being proposed (which require support) are the testing of a new synthetic biomimetic valve which could offer an alternative to the mechanical devices.

C	The **Character's** strength is that the team have already developed a product (with the support of the British Heart Foundation).
K	The **Key Benefits** are that the new synthetic valves made by GM's team have good mechanical properties which, if they also have longevity, would eliminate the need for patients to take warfarin.
L	The **Lure** is the statistical evidence of the scale of heart disease in the UK.
E	The **Ending** is GM asking for a meeting to discuss the proposition further.

This was a first Pitch which addressed each of the *CRACKLE* components. Whether the details are completely accurate is not important in the first draft. They can easily be checked and changed at a later stage. Initially there are almost certainly big gaps where information on Key Benefits or relevant Reasons to Care are unknown or unclear. The framework helps to focus attention on each *CRACKLE* component. It acts as a prompt to do the research to make the Pitch proposal as persuasive as possible.

In the above example, the next stage would be to drill down into each ingredient. Are the Reasons to Care significant enough? Are there more which could strengthen the Pitch? How many younger people need new heart valves? What is the size of the potential market? In the Key Benefits section, how much would be saved in health costs if regular checks for warfarin were no longer needed? Would it be cost effective for older people to be given the new valve?

Looking more broadly at the Actions section, is there any evidence that there are alternatives to mechanical devices as well as biomimetic devices; perhaps some new scientific development which could make both redundant?

These are some questions which arise through simple curiosity from the first draft of the Pitch. They, and others which arise at each iteration, can help the pitcher make the elevator Pitch stronger and clearer. It can also help with the full detailed proposal.

An example of a Pitch to a potential partner

Simon Rayner and Elliot Gold are serial entrepreneurs. They notice where people struggle and stumble, and then apply themselves to designing profitable solutions. One problem they noticed was the difficulty that many busy people experience when they try to fix appointments using their mobile phones. It is tricky to co-ordinate diaries when you are on the move. Their Challenge was to create an App to make it easy to arrange mutually convenient appointments on mobile phones. These are the *CRACKLE* elements which they used to tailor their Pitch to a potential chief technical officer, someone they needed to help turn their idea into a marketable product.

C The **Challenge** is to develop the App, SyMeet – a kind of Doodle for mobiles.

R The **Reasons to Care** are:

Millions struggle to make mobile appointments. They are difficult because mobile calendar interfaces are fiddly and people are often unable to see their calendar while they are on the phone. It can take several email/text exchanges to agree a mutually convenient appointment time, causing delays and frustration.

Although solutions like Google Hangout Requests, Doodle and WhenIsGood exist, they are either limited in scope or are web applications which struggle to migrate smoothly to a mobile platform. Think of how difficult Skype's transition to mobile has been (for years slow, cumbersome and still requiring Skype IDs). Compare that to how quickly Viber or WhatsApp have grown as native-mobile applications: intuitive and easy to use. We believe SyMeet, with your help, could be similarly successful.

A Our **Actions** to date: we have already created a working system for a simple, cross-platform, mobile application that makes the whole process of making appointments on mobile phones very simple and straightforward – even if only one of the parties has the App. Our goal is to work towards acquisition – once SyMeet is established and has strong viral growth metrics.

C The **Characters**: Simon has 18 years experience in IT. He founded, grew and sold a technology company. Elliot is an ex-PricewaterhouseCoopers management consultant with experience in corporate and start-up worlds.

K The **Key Benefits** are that SyMeet is already published on the Google Play Store and we have a clear development plan which we can share with you. But there are also opportunities for you to help us shape the plan and refine the offer in the market place.

L The **Lure**: An opportunity to become a partner in our tech start-up company, based at Silicon Roundabout in London's Tech City, and to develop what we believe will be an essential App. An App which the giant IT companies – Google, Microsoft etc. who are vying for market share – may well want to acquire.

E **Ending**: Our App solves a genuine problem. We are a team with a clear vision and track record. We would like to book 30 minutes to demonstrate the App, show you the technology we are using and discuss how you might help us going forward, and what we can offer you in return.

An example of a summary for an impact statement

And finally, this is a contribution from Dr Joanne Heng from Glasgow University. She used the CRACK of the *CRACKLE* format to draw together the information she needed to write a 150-word summary of her research.

> There is an urgent need to develop new drugs and vaccines to prevent and treat malaria, a parasitic disease which is a major public health concern. *Plasmodium*, the parasite responsible for malaria, has a close relative, *Toxoplasma gondii* (responsible for toxoplasmosis). *Toxoplasma* is a more convenient model to study than *Plasmodium* due to the ease with which it can be genetically manipulated.
>
> One approach in drug and vaccine development against malaria is to study parasite invasion mechanisms in *Toxoplasma*. Our laboratory has recently developed an efficient system for deleting individual genes. We can therefore rapidly generate a range of mutant *Toxoplasma* parasites which lack the necessary proteins for invading their host. Essential proteins identified in *Toxoplasma* which are also found to be important in *Plasmodium* survival are likely to provide suitable targets for drugs which can be developed to prevent the spread of malaria.

C The **Challenge**:

To identify the protein components of the parasite *Toxoplasma gondii*, which are necessary for successful host invasion.

R The **Reasons to Care** are:

Toxoplasma gondii is a close relative of another parasite, *Plasmodium*, which is responsible for malaria.

1. Malaria is one of the biggest killers in the world. More than 200 million cases are reported annually – almost half of those infected die.
2. Being able to identify the components (proteins) involved when *Toxoplasma gondii* invades its host could be a first step in identifying which of those may be involved when *Plasmodium* invades its host.
3. Those proteins which are found to be necessary for *Plasmodium* to invade its host, and are also essential for its survival, are likely to be suitable targets for drug development.

A The **Actions** being proposed:

1. Delete specific genes (thought to be involved in the invasion process) to produce a range of mutant *Toxoplasma* parasites and investigate the consequences of each gene deletion. (That is, are they still able to invade their host?)
2. Characterise the various mutants to identify which components of the invasion apparatus the parasite needs to survive.
3. See if the components necessary in *Toxoplasma* survival are the same as in *Plasmodium.*
4. Those that are can act as models with which to target *Plasmodium* components with drugs.

C The **Character** is the laboratory team.

Our team has expertise in the molecular and cellular biology of *Toxoplasma gondii*. We have developed cutting edge technology to efficiently delete genes enabling the rapid development of mutant parasites.

K The **Key Benefits** are:

We can achieve an insight into parasite invasion mechanisms to aid in the development of drugs and vaccines to prevent the spread of malaria.

Specifically:

Identification of the relevant 'survival' proteins in *Toxoplasma gondii* and its relative, *Plasmodium*, will make it possible to begin to design drugs and vaccines targeted to those specific proteins in Plasmodium. In other words, information gleaned from *Toxoplasma gondii,* which is relatively easy to manipulate genetically, can serve as a model for understanding *Plasmodium* in which genetic manipulation is more difficult.

References

Introduction

1. Faraday, M. (1870). 'Advice to a Lecturer', in Jones, B.H. (ed.) (1960), *Life and Letters of Michael Faraday*, Royal Institution of Great Britain, London (p.135).
2. McKee, R. (1997). *Story, Substance, Structure, Style and the Principles of Screenwriting*, Methuen, London.
3. Booker, C. (2004). *The Seven Basic Plots*, Continuum, London.
4. Simmons, A. (2001). *The Story Factor, Inspiration, Influence, and Persuasion through the Art of Storytelling*, Basic Books, New York.
5. Denning, S. (2001). *The Springboard, How Storytelling Ignites Action in Knowledge-Era Organizations*, Elsevier, Amsterdam. See also: Denning, S. (2005). *The Leader's Guide to Storytelling, Mastering the Art and Discipline of Business Narrative*, John Wiley and Sons, San Francisco.
6. Lawson-Tancred, H. (trans.) (1991). *Aristotle: The Art of Rhetoric*, Penguin Books, London.

1. Your Audience

1. Michael Faraday cited in Day, P. (ed.) (1999). *The Philosopher's Tree, Michael Faraday's Life and Works in his own Words: A*

Selection of Michael Faraday's Writings, IOP Publishing Ltd, Bristol and Philadelphia.
2. Maslow, A.H. (1954). *Motivation and Personality*, Harper, New York.
3. van den Brul, C. (1987). *QED, Eating Earth*, BBC One, London, TX 22/04/1987.
4. Smith, J.A. (trans.) (2006). *Aristotle: On the Soul*, Digireads, Stilwell, KS.
5. King, S. (2000). *On Writing*, Scribner, New York.
6. Schank, R.C. (1990). *Tell Me a Story: Narrative and Intelligence*, Northwestern University Press, Evanston.
7. This theory is elaborated in Gardner, H. and Laskin, E. (1995). *Leading Minds: An Anatomy of Leadership*, Basic Books, New York.
8. Reported in Thomas, J. (2009). *Practising Science Communication in the Information Age, Theorising Professional Practices*, Oxford University Press, Oxford. For the section on controversy and consensus, see pp. 113–148.
9. Reported in Harris, L. (2012). The simple truth, *Wellcome News*, Issue 71, Spring, pp. 14–19.

2. Change and Affect

1. Chief of Staff Rahm Emanuel and President Obama in the Oval Office. (Photo: Pete Souza/White House, USA.)
2. Draft figures 1–4 of the second Watson and Crick paper, published as Crick, H.C. and Watson, J.D. (1953). Genetical implications of the structure of deoxyribonucleic acid, *Nature*, 171(4361), 964–967. Photo: Wellcome Library, London.
3. Watson, J.D. (1968). *The Double Helix*, Atheneum, New York.
4. Tolstoy, L. (1869). *War and Peace*, Maude, L. and Maude, A. (trans.) (1943), Macmillan, London (pp. 473–474).

3. Lure

1. Leadership Academy Workshop (2007). *Leadership – Innovation and Inspiration*, Royal Holloway, London, 14 June 2007.

2. Anon (1994). Front page, *Daily Star*, 25 May 1994. www.expresspictures.com
3. Photographs courtesy of Médecins Sans Frontières and TW CAT, Brighton.
4. Lewycka, M. (2005). *A Short History of Tractors in Ukrainian*, Penguin Books, London. *A Short History of Tractors in Ukrainian* by Marina Lewycka (Copyright © Marina Lewycka, 2005). Reprinted by permission of A.M. Heath & Co Ltd.
5. Hershberg, J.G. (1993). *James B. Conant: Harvard to Hiroshima and the Making of the Nuclear Age*, Knopf, New York.
6. British Science Festival 14–19 September 2010, Birmingham, England.
7. Geim, A.K. (2012). *The Frog That Learned to Fly*. Available online at: www.ru.nl/hfml/about-hfml/levitation/diamagnetic/ [Accessed 11 November 2012].

4. World

1. Bransford, J.D. and Johnson, M.K. (1972). Contextual prerequisites for understanding: Some investigations of comprehension and recall, *Journal of Verbal Learning and Verbal Behaviour*, 11(6), 717–726.
2. The National Commission on Terrorist Attacks Upon the United States (2004). *Final Report of the National Commission on Terrorist Attacks upon the United States*. Available online at: http://www.9-11commission.gov [Accessed 11 November 2012].
3. Jacobson, S. and Colón, E. (2006). *9/11, The Illustrated 9/11 Commission Report – The full story of 9/11 before, during and after the attacks*, Hill and Wang, New York; Viking, London.
4. Foreword, Copyright 2006 by Thomas H. Kean and Lee H. Hamilton.
5. Higgs Boson: One page explanations, *physics4me*. Available online at: http://physicsforme.wordpress.com/2011/07/24/higgs-boson-one-page-explanation/ [Accessed 25 July 2013].
6. Miller, D.J. (1993). *A Quasi-political Explanation of the Higgs Boson; for Mr Waldegrave, UK Science Minister 1993*. Available

online at: http://www.hep.ucl.ac.uk/~djm/higgsa.html [Accessed 6 August 2012].

7. Copyright of Gigerenzer, G. (2002). *Reckoning with Risk, Learning to Live with Uncertainty*, Penguin Books, London (pp. 5–6).
8. Slovic, P., Monahan, J. and MacGregor, D.G. (2000). Violence risk assessment and risk communication: The effects of using actual cases, providing instruction, and employing the probability versus frequency formats, *Law and Human Behavior*, 24, 271–296.
9. Reported in Kahneman, D. (2011). *Thinking, Fast and Slow*, Farrah, Straus and Giroux, New York (p. 330).
10. Thomas, B. (1976). *Walt Disney: An American Original*, The Walt Disney Company, New York (p. 246).

5. Character

1. Perutz, M. (1989). *Is Science Necessary? Essays on Science and Scientists*, Barrie and Jenkins, London (p. xiv).
2. Holmes, O.W. (1891). 'Scholastic and Bedside Teaching, Introductory Lecture to the Medical Class of Harvard University (6 November 1867)', in Holmes, O.W., *Medical Essays 1842–1882* (p. 302). Available online for download at: http://onlinebooks.library.upenn.edu/webbin/gutbook/lookup?num=2700 [Accessed 30 July 2013].
3. Glasman-Deal, H. (2010). *Science Research Writing for Non-Native Speakers of English*, Imperial College Press, London. Holtom, D. and Fisher, E. (1999). *Enjoy Writing your Science Thesis or Dissertation*, World Scientific, Singapore.
4. Kane, T.S. (1988). *The New Oxford Guide to Writing*, Oxford University Press, Oxford. King, G. (2009). *Collins Dictionary Publishing, Improve your Writing Skills*, Harper Collins, London.
5. Copyright 2012, Plain English Campaign. Plain English Campaign (2012). *Golden Bull Awards.* Available online at: http://www.plainenglish.co.uk/awards/golden-bull-awards/golden-bull-winners-2012.html [Accessed 5 April 2013].

6. Hope, C. (2012). Alan Duncan issues memo at DFID banning jargon words such as 'going forward', *Daily Telegraph*, 22 June 2012. Available online at: http://www.telegraph.co.uk/news/politics/9349970/Alan-Duncan-issues-memo-at-DFID-banning-jargon-words-like-going-forward.html [Accessed 30 July 2013]. Original memo also received: Personal communication, 2013.
7. Gregory, M.W. (1992). The infectiousness of pompous prose, *Nature*, 360, 11–12.
8. The original article containing this quote is Rowley, D. and La Brody, J.J. (1986). *Journal of Diarrhoeal Diseases Research*, 4, 1–9.
9. Austin, J., Butchart, N. and Shine, K.P. (1992). *Nature*, 360, 19 November, 221.
10. Rowling, J.K. (1998). *Harry Potter and the Chambers of Secrets*, Scholastic Inc, London (p. 333).
11. Slee, P. (2011). *Engage: Understanding Strategy and Making it Work?*, Autumn 2011, Leadership Foundation for Higher Education, London.
12. Salinsky, T. and Frances-White, D. (2008). *The Improv Handbook: The Ultimate Guide to Improvising in Comedy, Theatre, and Beyond*, Continuum International Publishing Group, London.
13. Johnstone, K. (1979). *Impro: Improvisation and the Theatre*, Faber and Faber, London.
14. The Spontaneity Shop workshops. Information available online at: http://www.the-spontaneity-shop.com/workshops/ [Accessed 27 July 2013].

6. Big Hook

1. Reported in *12 Timeless Writing Tips from Mark Twain*. Available online at http://www.onlineuniversities.com/blog/2010/07/12-timeless-writing-tips-from-mark-twain/ [Accessed 24 July 2013].
2. President John F. Kennedy, September 12 1962, Rice University, Houston, Texas. Available online at http://www.youtube.com/watch?v=G_7K0yIwN4E [Accessed 11 November 2012].

3. Kahneman, D. (2011). *Thinking, Fast and Slow*, Farrah, Straus and Giroux, New York (p. 32).
4. Hess, E.H. (1965). Attitude and Pupil Size, *Scientific American*, 212, 46–54.
5. Sykes, C. (1981). *Horizon: Pleasure of Finding Things Out*, BBC Two. Broadcast on 23 November 1981. Available online at http://www.youtube.com/watch?v=Bgaw9qe7DEE [Accessed 31 July 2013].
6. Lynch, J. (1995). *Horizon: Fermat's Last Theorem*, BBC Two. Broadcast on 3 December 1995. Available online at http://www.youtube.com/watch?v=7FnXgprKgSE [Accessed 25 July 2013].
7. Widely attributed to Einstein. Reported in Spradlin, D. (2012). Are you solving the right problem?, *Harvard Business Review*, September 2012. Available online at http://hbr.org/2012/09/are-you-solving-the-right-problem/ [Accessed 30 July 2013].

7. Plot

1. Voltaire (1738). *Sixième discours: Sur la nature de l'homme, Sept Discours en Vers sur l'Homme*. Available online at: http://en.wikiquote.org/wiki/Voltaire [Accessed 25 July 2013].
2. Booker, C. (2004). *The Seven Basic Plots*, Continuum, London and New York.
3. Medawar, P. B. (1979). *Advice to a Young Scientist*, Basic Books, New York (p. 3).
4. Butcher, S.H. (trans.) (1978). *Poetics of Aristotle XVII*. Available online at: http://www.gutenberg.org/files/1974/1974-h/1974-h.htm [Accessed 11 November 2012].

8. Structure

1. Faraday, M. (1870). 'Advice to a Lecturer', in Jones, B.H. (ed.) (1960), *Life and Letters of Michael Faraday*, Royal Institution of Great Britain, London (pp. 1–8).
2. Banyai, I. (1995). *Zoom*, Viking Children's Books, New York.

3. Reported in Dutton, K. (2010). *Flipnosis, The Art of Split-Second Persuasion*, William Heinemann, London.
4. Stanford Report (2005). *Text of Steve Jobs Stanford Commencement Address*. Available online at: http://news.stanford.edu/news/2005/june15/jobs-061505.html [Accessed 6 April 2013].
5. Butcher, S.H. (trans.) (1978). *Poetics of Aristotle XXIII*. Available online at: http://www.gutenberg.org/files/1974/1974-h/1974-h.htm [Accessed 11 November 2012].
6. Faraday, M. 'Letter to B. Abbot (31 December 1816)', in Day, P. (ed.) (1999), *The Philosopher's Tree: Michael Faraday's Life and Work in his own words*, Institute of Physics Publishing, Bristol and Philadelphia.

9. Make your Pitch *CRACKLE*

1. Roberts, W.R. (trans.) (1954). Rhetorica: The Works of Aristotle, XXX.i. Available online at http://rhetoric.eserver.org/aristotle/rhet3-1.html [Accessed 11 November 2012].
2. Twain, M. (1876; reprinted 2006). *The Adventures of Tom Sawyer*, Penguin Books, London (p. 23).
3. REF (2010). *Research Excellence Framework: Impact pilot exercise*. Available online at: http://www.ref.ac.uk/media/ref/content/background/impact/EarthSystems_EnvironmentalSciences.pdf [Accessed 3 June 2012]
4. Photograph of biomimetic heart valve courtesy of Dr G. Moggridge.

Index

Printed in Great Britain
by Amazon